20th Century Bioscience

Professor O. J. Eigsti and the Seedless Watermelon

In appreciation to Dr. Gene Stutzman and associates
O. J. Eigsti
11-9-99

20th Century Bioscience

Professor O. J. Eigsti and the Seedless Watermelon

John H. Woodburn, Ph.D

Pentland Press, Inc.
England • USA • Scotland

Cover photograph provided with permission from the Goshen Hospital and Healthcare Foundation, Inc., Goshen, Indiana. Photo by Verlo Yoder, Goshen, Indiana.

PUBLISHED BY PENTLAND PRESS, INC.
5122 Bur Oak Circle, Raleigh, North Carolina 27612
United States of America
(919)782-0281

ISBN: 1-57197-132-7
Library of Congress Catalog Card Number 98-067426

Also by John H. Woodburn:

Opportunities in Chemistry Careers. National Textbook Co., 1997, First edition, 1971.

Opportunities in Energy Careers. National Textbook Co., 1987, First edition, 1978.

Scientists and Science Teachers. Genie Publications, 1990.

Taking Things Apart and Putting Things Together. American Chemical Society, 1976.

The Whole Earth Energy Crisis. G. P. Putnam's Sons, 1973.

Demonstrations and Activities for High School Chemistry, with L. R. Sanford and W. J. Brawley Jr., Parker, 1971.

Know Your Skin. G. P. Putnam's Sons, 1967.

Excursions into Chemistry, Lippincott, 1965.

Cancer: The Search for its Origins. Holt, Rinehart, and Winston, 1964.

Teaching the Pursuit of Science, with Ellsworth S. Obourn. Macmillan, 1964.

Radioisotopes. Lippincott, 1962.

General Science Boy Scout Merit Badge Booklet. First published in 1972.

Energy Boy Scout Merit Badge Booklet. First published in 1978.

To our forebears who taught us to revere Nature
and to love the land,
And those whom we taught and who, in turn,
taught us.

Table of Contents

Preface

THIS IS THE life story of only one of the millions of scientists, teachers, and agriculturists who will have played out their twentieth century lives adding to and taking from the streams of consciousness that make life worthwhile. This is the life story of Dr. Orie J. Eigsti. At the same time, the story I will tell is about much, much more. It is about what it means to be a scientist, teacher, horticulturist, husband, father, friend—any one or all of us who are trying to make our lives count for something in this intellectually fertile century. This is a wonderful way to home in on the evolution of science, education, and agriculture and see how this evolution creates opportunities to go where we want to and do what we want to do.

The urge to see the seedless watermelon become a commercial success is relentless; this urge has taken O. J. on worldwide missions, consumed hundreds of thousands of dollars, and plunged him into the depths of discouragement only to lift him eventually to the heights of success. In a very real sense, the teaching profession made it all possible. Stretching as it does across most of the twentieth century, O. J. Eigsti's life story catches the spirit of using the past to mirror the future. The hopes and premonitions of youth intermingle with the realities of those who have been there, and better appreciate each other by doing so.

The story I will tell gains its value from the continuity of the human endeavors in which O. J. has been engaged. In science, humanity's struggle to make nature and people good for each other and coexist goes on and on. In agriculture, nothing is more critical and ongoing than providing food and fiber for the world's people. And what greater contribution to the advancement of civilization can one make than to improve the education of each new generation?

O. J. and the modern pursuit of science have grown up together, shared adolescence, faced mid-life crises, and met the responsibilities and shared the satisfactions that come with maturity. There are lessons here that are not permitted in the crowded pages of science textbooks. There is much to learn about education by joining the students in all eight grades of a one room school, "dropping out" after high school, but nine years later, receiving the final blessing of one's doctoral committee. There is much to learn about farming by trudging behind a horse-drawn cultivator along seemingly endless rows of corn and ultimately bidding god-speed to the drivers of eighteen-wheelers departing on cross-country deliveries of seedless watermelons. It is much more comfortable to let O. J. tell us what it is like to train to be a teacher, only to be denied a teaching position during the Great Depression, but to ultimately hold full professorial rank at highly respected universities.

The life story of O. J. Eigsti is not a textbook, but it is what textbooks try to teach. As such, it is an invitation opening the way to where textbooks would have us go. It is any one of us wanting to share the ups and downs, the highs and lows of what it means to be a scientist, farmer, and teacher, each vocation with its own motivations, hopes, and ambitions. At the same time, each is shown as nurturing and depending on the others, and all the while defining what it means to be a person.

I would be severely remiss were I to delay giving credit to the authors of a book that gives me the courage to attempt to translate the life of a friend into mere words. In their *Generations: The History of America's Future* (1991), William Strauss and Neil Howe describe eighteen generations of Americans whose lives illustrate the entire American drama. This book provides a kind of infrastructure for contemplating not only the life of O. J. Eigsti, but of each of us.

Such an infrastructure, however, provides perspective only. Strauss and Howe include "sample members" of each of the eighteen generations, but the lives of O. J. Eigsti and the vast segment of America he represents receive little attention. Rarely are scientists or teachers included. The millions of Americans in each generation who produce food, fiber, and other horticultural and agricultural products receive even less recognition, although these products sustain our lives and make possible large segments of the American economy.

Eigsti's generation was the GIs, those born between 1901 and 1924; as of 1996, this group includes seven presidents and other such notables as Charles Lindbergh, Bob Hope, Katherine Hepburn, Joe DiMaggio, Robert McNamara, Walter Kronkite, Billy Graham, Sidney Poitier, and Lee Iacocca. The only scientist included among these "sample members" is

Robert Oppenheimer; no educators or agriculturists appear in the sample member list.

The "Missionary" generation, those born between 1860 and 1882, included four presidents and other such "sample" members as Jane Addams, Billy Sunday, Henry Ford, Frank Lloyd Wright, Orville Wright, Isadora Duncan, Margaret Sanger, Douglas MacArthur, H. L. Mencken, Helen Keller, and George C. Marshall. Again there is only one scientist listed, Albert Einstein, and no educators or agriculturists. This is O. J.'s parents' generation and the one that would be pretty much "tending the store" while Eigsti was establishing a place for himself in the professional world.

Again, from the point of view of "tending the store" while Eigsti was ending his most productive years, the "Silent" generation was those born between 1925 and 1942. Marilyn Monroe, Sandra Day O'Connor, Martin Luther King Jr., T. Boone Pickens Jr., James Dean, Woody Allen, Ted Koppel, and Pat Schroeder are cited as "sample members" of this generation. No presidents, scientists, educators, or agriculturists were listed.

The Eigstis' sons, Karl and Nick, by dates of birth are members of this "Silent" generation; but only a few years kept them from being "boomers," that widely discussed generation whose birthdates fall between 1943 and 1960. In fact, first as parents of children who became of age during the 1960s and were the dominant generation during the final quarter of the twentieth century, the lives of O. J. and Agnes Eigsti may well have been influenced as much by their sons' generation as by their own. For sure, both as a parent and a teacher of the boomers, interactions between two powerful generations, the GIs and the boomers, make the life story of O. J. Eigsti a unique opportunity to revisit the twentieth century.

As of 1997, the generation that is represented by the Eigstis' sons produced one president, but, again, no scientists or agriculturists were cited as sample members. Significant events, however, did include the discovery of polio vaccine, the exploration of space, and science education becoming a national priority. The sample members cited (1991) include Oliver North, Joe Namath, Donald Trump, Jane Pauley, Oprah Winfrey, Bill Gates, and John McEnroe. Political leaders cited include Bill Bradley, William Bennett, Newt Gingrich, Dan Quayle, and Albert Gore, Jr.

The analyses of how each generation has influenced and continues to influence our society have been crucial in selecting the events and circumstances that most probably played significant roles in shaping the life of O. J. Eigsti and in establishing the dimensions of his impact as a scientist and educator. Actually, there are advantages in looking at the century-long and nationwide evolution of things by focusing on the life of only one person. It is good to know how our society affects and is affected by societal changes, but there is no escape from being one's own person.

And that is what O. J. Eigsti is and has been. The central focus of his life will always be on the seedless watermelon. But his total life story includes unique opportunities to see how sweeping social changes influence and are influenced by farmers, scientists, and teachers, particularly those who have inherited not only the wherewithal but the motivations, ethics, and lifestyle that distinguish rural Americans from their urban compatriots.

For me, this is a new kind of writing with new challenges. In the past, it has been relatively easy to describe how scientists come up with their hypotheses, to trace their origins to the gradual accumulation of knowledge and know-how. The strategy of their investigations, the design of their experiments, and the significance of the results are readily extracted from the available literature. But creating a valid word picture of a life is a very different challenge. At best, it is creating a jigsaw puzzle for readers to assemble. The things O. J. has done and the places he has been become the foreground, the part of the puzzle that is easy to assemble. The century-long evolution of genetics, agriculture, and education form the background, and as in real puzzles with cloudy skies or featureless seascapes, there is much trouble in finding which piece goes where. Nor is it easy to fashion or to find the pieces that join foreground to background.

Unlike jigsaw puzzles, there is no printed picture of the assembled puzzle to serve as preview. Each reader is free to enjoy the sense of discovery, to decide for themselves who O. J. Eigsti is and what he stands for. If there are pieces missing or bent out of shape in trying to force them into places where they don't really belong, I can only apologize.

Appreciation of the many citations from publications and personal correspondence is hereby expressed and their value acknowledged. Those not mentioned in the text include the editorial staff of the *Cucurbit Network News*, Cara Dewey at the National Watermelon Association, Dr. Laurie Hodges at the University of Nebraska-Lincoln's Cooperative Extension, Linda McElfresh at Agricultural Research, and Claude E. Thomas at the U.S. Vegetable Laboratory.

I am uniquely indebted to those in the Leisure World Writers Group, who critiqued working versions of much of this story. Similar help came from the Fossils, a retired men's organization, particularly Glennon Loyd and Alfonso Geiger, who shared with me their extensive collections of other authors' efforts to keep alive the challenges and satisfactions that are exemplified in the life of O. J. Eigsti.

Special thanks are due to the Goshen Hospital and Health Care Foundation for permission to use the August page of their 1997 calendar for the jacket design. Hope Pantell, Mrs. Ruth Woodburn and Mrs. Charlene Monk provided essential and much appreciated editorial assistance.

Chapter 1
A Scientist, Educator, Horticulturist and Friend

INTRODUCTIONS ARE WONDERFULLY important. They are the first step toward knowing someone to turn to for advice or assistance. And introductions that lead to new friendships are doubly valuable. This is what I want to happen as you meet Dr. Orie J. Eigsti.

The time is early spring of 1949, right in the middle of the twentieth century. The place is a greenhouse on the campus of Illinois State Normal University. You cannot shake hands because the man I want you to meet is carrying large wooden trays filled with watermelon seedlings. Nor is there time for small talk. Both the temper and the actions of the man we have just met tell us there is always something very important to do.

Each morning for the next five days we stop by the greenhouse and perform a curious ritual that begins with putting a tuft of cotton between the seed leaves of each watermelon seedling. Then, we add a few drops of a clear solution to the cotton. The solution contains colchicine, which we are told is a widely recognized drug obtained from the autumn crocus. There is only time for brief mention of chromosomes, diploid, tetraploid, along with confusing references to watermelons without seeds. Then, the man we just met hurried along to his job as geneticist at the Funk Brothers Seed Company.

The man was O. J., the scientist. For the next five days, we do our assigned chore, not so much because we know the whys and wherefores of what he asked us to do, but because we have caught a bit of his natural enthusiasm and with it, the tacit assurance that something important will happen to those watermelon seedlings. What happened is a large part of O. J. Eigsti's life as it has unfolded throughout the twentieth century.

Actually, meeting O. J. in the homely surroundings of a greenhouse tells us only a little of who he is, where he has been, what he has done, and what he continues to be hell-bent on doing. His side of our conversations are far more revealing than what I can tell you. The following examples begin with a reference to O. J. attending a professional meeting:

> *I went to the International Genetics Conference in Stockholm. The Genetics Society of America gave me a travel grant. I was at Northwestern [University] at the time. This was 1948 and my first trip abroad. I met scientists that I had read about in my graduate studies: Winge, Tschermak, Haldane, Fisher, and others. I heard several papers by Professor Kihara. He was given special permission by General MacArthur to attend as a representative of Japan since Japan was not yet out of the war/peace phase. Of most interest to me was the work done involving colchicine and the most surprising and interesting application was the production of tetraploid/triploid watermelons whereby a seedless fruit was produced.*
>
> *Before I left for Sweden, a neighbor across the street in Evanston had asked me to look up a cousin, Siegfrid Edstrom, in Stockholm. When I could get to it, I phoned and his secretary asked if I knew him. I said no. Well, it turned out he was very wealthy and head of the Telephone Company in Sweden, which was equivalent to [being] the president of General Electric! He told me that he knew I was coming to Sweden and he wanted to have a dinner for me. The dinner was a fine affair with service I had never experienced before.*
>
> *It was the other guests who made this a most memorable occasion. Professor Muntsing was invited because he was heading up genetic research in Sweden. It turned out that Edstrom was a very important person in the Olympics and Professor Kihara was a star baseball player, a member of Japan's Olympic ski team, and a geneticist. So here was I sitting there and Kihara started telling me what he was doing with the development of seedless watermelons in Japan. And he knew I was working on a book covering the use of colchicine to produce tetraploids. We became friends immediately. I invited him to stop off and visit us in Evanston on his way back to Japan. Before I left Stockholm, he gave me a roll of kroners [Swedish crowns] to take back home so he would have more money to spend in the USA. Quite a trust to say the least!*
>
> *So, in August, Kihara came to Chicago. We had him to lunch. On the second day, I said I am sorry but I promised Karl we would go to the baseball game. [He said,] "Oh! I'd love to go. I went yesterday." But it was not without receiving questioning looks from others who were still thinking of Japan as our mortal enemy.*

> *So here I am, working on a book dealing with colchicine and visiting with someone who is way out front in research on its practical uses. I said, "Why don't you write up what you are doing and have it published in the U.S.? This is very interesting and there are a lot of watermelon people who would be very interested." Now comes 1951. I got this letter from him and a manuscript on hybrid triploid watermelons. It was a very terse letter. I could read between the lines [that] he was very unhappy because the manuscript had been rejected.*
>
> *I was about to leave for a trip to Long Island for a meeting so I stopped in New York to see the editor of the Proceedings of the American Horticultural Society. I told him, "Here is a manuscript that should be published." He looked at it and leaned back and kinda smiled and said, "We had that and we turned it down." I said, "You did!" "Yes, it was sent back. Here are the reviewers' letters." You know the animosity between those people and Japan at war had so colored their judgment. And they didn't really understand what the paper was all about because it takes two pollinations to make seed and these reviewers couldn't understand this. I said, "This is terrible. This is perfectly good science."*
>
> *"Well" he said, "we need a little money. We need $75 to publish the paper." "Well," I said, "that's no problem. I'll just ask five seed companies to come up with $15 each." One seed company refused, so I put in the $15 myself. That company had been the first to get seedless melon seed from Japan and it had been a totally miserable failure. So what happens to this paper? A year later I was in Pakistan and I got a letter saying [I would be] interested to know that Kihara's paper won the prize for being the best publication in 1951. So now Kihara becomes tremendously grateful to me, and in 1953, when we met at a conference in Italy, he came to me and said, "Guess what happened." I said I [didn't] know. He said, "I won the prize for having the best paper."*

These quotations tell us much about the man I am introducing. Looking between the lines, we catch glimpses of personal attributes making the most of random coincidences. But there is no waiting for pregnant coincidences to fall in his lap. For more than sixty years, O. J. has traveled near and far, always looking for those rare combinations of circumstances that can be followed up and make life worthwhile both for himself and for others.

Another introduction to O. J. Eigsti occurs at the Presbyterian Church quite close to the campus of Illinois State Normal University. The minister has invited anyone who was so inclined to serve as guest preacher. Many of us may have fantasized about preaching a truly soul-saving sermon, but this morning we meet someone who sees little reason to forgo the possible satisfactions that come with translating fantasy into reality. I didn't know what to expect when O. J. went to the pulpit. I knew that the Eigstis had grown up in the Mennonite faith, but I didn't associate Agnes Eigsti with a little white cap or O. J. with somberness in costume or spirit.

As is traditional in most churches, O. J. read the text on which his sermon would be based: "If thou hast run with the footmen, and they have wearied thee, then how canst thou contend with the horses? and if in the land of peace, wherein thou trustedst, they wearied thee, then how wilt thou do in the swelling of Jordan?" (Jeremiah 12:5)

To be honest, even after asking for clarification, the point of his sermon escaped me. I assume it came from within a part of O. J. that I never felt I should invade. But to introduce him without letting you know it is there would be misleading. Some four decades later, here is the point of the sermon as O. J. remembered it:

> *The point of the sermon was that as you grow older life doesn't get any easier and if you're in grade school and you think life is difficult, wait until you get to high school. And if you think high school is difficult, wait until you get to college. If you think college is difficult, wait until you get to graduate work, and if you think graduate work is difficult, wait until you get to be an intern at the Carnegie Institution and you'll find out what difficult means. And if you can't keep up in that stream, you're going to be in a bad way. That's the whole point. But by the grace of God we can do it. There's something outside of us that keeps us going. Some stimulus, whether it is a family relationship, or friends, or just like it was when the Larsons, my very first growers, who came to me and said, "Now don't you give up on seedless watermelons." In other words, I was ready to quit running with the footmen and Larson said, "no, you can't give up, you must keep running." That's the moral of that sermon.*

The "ceremony of the woods," with its almost religious motif, suggests other attributes of O. J. Eigsti. The ceremony usually begins on a Sunday morning with Agnes packing a picnic basket with the provisions to prepare and serve breakfast in the middle of a twenty-acre wood that is an amazingly true representation of the forests that were once the environment that is now the Indiana countryside. O. J.'s role in the preparation chores is pretty much limited to packing a small traveling bag with a bottle of special brandy and the appropriate glassware.

With solemnity approaching that with which one enters a cathedral, we find a special area deep among spreading beech and other hardwood trees. On the way, we acknowledge an enormous granite boulder, a souvenir of nature's infinite forces acting by way of that huge glacier that did so much to make the nation's "breadbasket" what it is. With a reverence that suggests the Last Supper, O. J. serves the brandy. The toast that is raised with our glasses is our appreciation of still being able to experience nature very much as it was before humanity's needs and interests took over. But, here too, we feel the premonition that tomorrow is fraught with desecration.

Later, we survey the undergrowth with its array of jack-in-the pulpits, triliums, May apples and their early spring compatriots, and the hundred-year-old walnut, ash, wild cherry and other species enormously valuable for the lumber they would yield. It is natural to ask O. J. how was he fortunate enough to hold dominion over such a vanishing resource. His simple answer describes both who O. J. is and the Eigsti attitude toward money.

> *I wanted a beech woods. For years I had collected Polygonatum for pollen tube slides in this woods. When I heard it was for sale. we bought it. We had no money so we got a mortgage from the bank that would require $350 per month for ten years. Since I had a tenured teaching job, I thought we could handle this.*

To continue our introduction to O. J. Eigsti, join my class in science for the non–science majors at the Illinois State Normal University. O. J. is to be a guest lecturer; his topic, the seedless watermelon. You will find his lecture well-organized and presented equally well. Most of all, I remember the look on O. J.'s face when he reached the high point of his lecture. He was about to cut open a handsome specimen of the seedless melons he was developing. When not a single seed appeared as the two split halves cracked open, the look on his face changed immediately from fearful premonition to sheer exultation.

It was 1952 when this one melon left no doubt but that it was possible to produce seedless watermelons. Thirty-five years passed, however, before they would be commercially successful—thirty-five years of ups and downs that consumed the bulk of Eigsti's time, energy, and income, including $85,000 of borrowed money. Not until the late 1980s was the essential scientific knowledge be combined with the equally essential production know-how to yield financial rewards and the even greater satisfactions that come with having one's dreams come true.

Another way to meet O. J. Eigsti will always be open to everyone. Visit the library at a major university or research laboratory and look up *Colchicine—In Agriculture, Medicine, Biology, and Chemistry*. The authors are O. J. Eigsti and Pierre Dustin, Jr.; the publisher is Iowa State College Press.

The contents of this 470-page book, including 1,652 citations of colchicine-related research, do more than tell a story that began "in the myths and the written records of ancient Egypt, India, and Greece, and runs its course through the ages into the world of today." It also reflects O. J.'s obsession to share his knowledge and know-how even though doing so demanded paying the price with endless hours spent at the writer's craft and dealing with the frustrations that come with finding a suitable publisher.

His reactions to the sweet and sour rewards that come with being an author tell us much about O. J. (There will be more about this in chapter 6.) For now, I quote only his reaction when I told him that the NIH copy has many underlined sentences and marginal checks, like "wrong, see ref #31." (It is to be noted that reference 31 is written in the German language.)

> *Well! What is that? I've never heard anything like that. Wrong? What was wrong? I didn't know this. That's very interesting. In German? This is new to me. Well for heaven's sake! I am going to look into that.*

Three rapidly evolving professions—genetics, teaching, and agriculture—come together in the life of the person I am introducing. During the twentieth century, there have been extremely significant changes in each of these professions. For genetics, enormous advances have been made in the knowledge of how traits and characters are handed down from parents to their offspring. Similarly, the ability to manage, manipulate, and control the transmission of inheritable traits continues to increase at a phenomenal rate.

Changes in the teaching profession have been even more widely publicized. As we follow the life of O. J., it will be particularly interesting to see how the education provided by the elementary and secondary schools of rural Illinois early in the century differs from that of today, and how what it takes to earn a Ph.D. at the University of Illinois today differs from that of 1935. Dr. Eigsti's role as a college professor is not easy to describe. Keeping body and soul together while seeking to develop the seedless watermelon comes to mind; but this narrow focus could shortchange his academic integrity and his urge to find better ways to teach. Actually, his roles as developer of the seedless watermelon and as a genetics professor nurtured each other. His students never doubted the significance of what he taught them.

As for changes in agriculture, Eigsti's home community of Goshen, Indiana, is a study of stability in a sea of upheaval. In much of the so-called Corn Belt, the old-time pattern of small homesteads no longer exists. The homes and barns that have not already been bulldozed to make way for massive farm machinery have been converted to storage facilities. But there are exceptions. Agnes Weaver Eigsti can still tend the flowers and pick raspberries in the garden of the home where she was born, and the Weaver name still appears in the pattern of colored slates on the roof of one large barn.

Some 100 years ago, Agnes's father was the water-boy when that barn was raised in typical Mennonite fashion. Only memories remain of the more than 100 neighbors who sat down to the celebratory dinner that was a meaningful part of the one-day effort. But the barn remains and can be seen as a symbol of the motivations, values, and virtues that have kept the Eigstis firmly attached to the soil and the agricultural way of life. No matter how far away they were physically or how alien the circumstances to which they had to adjust, they never abandoned their rural heritage.

The water-boy for that barn-raising was Joseph H. (for Hoover, but Hiram on the 1877 birth certificate) Weaver. Joseph was scarcely twenty-one when his persona revealed characteristics that may well have been handed down to his daughter Agnes. He studied engineering at the Elkhart Institute (later replaced by Goshen College), but found there were not many opportunities for Mennonite engineers. Using his coming-of-age gift of $3,000, he went west and invested in the Buck Gulch Gold Mine in Sumpter, Oregon, in the Blue Mountains. After only a few years, the wooden sluice boxes were destroyed by fire and were never rebuilt. Some of the gold, however, survived. When Joseph took some gold nuggets and dust to the assay office in San Francisco, he had jewelry made—gold studs and a ring with "Oregon" spelled out in Buck Gulch gold.

Circumstances brought Agnes's mother's family from Oregon to the farm in Goshen. A family story tells of daughter Agnes being born on a table board by the light of a kerosene lamp and five weeks later being carried across the threshold of the Administration Building of Goshen College to be officially welcomed into the Cradle Roll. The congregation of the College Mennonite Church held their services in the auditorium. Her mother was a charter member of the Church and attended the ground-breaking Ceremony for Goshen College in 1903. With little money, but a lot of hard work, her mother maintained high educational standards for Agnes and her nine brothers and sisters.

This introduction to the Eigstis would not be complete without quoting the motto that O. J. has repeated so many times:

> *If you have never failed, it is an even chance that you have never won a great success. I tried to get a job as a high school teacher and coach. I failed, so I enrolled in graduate work at the University of Illinois. After earning an M.A. in botany, I failed to get a teaching job so I went for the Ph.D. Even with a Ph.D. in botany, again no teaching job was available so I accepted work as a research assistant at the Carnegie Institute of Washington laboratory at Cold Spring Harbor. This didn't work out to my advantage but a job opened up for me at the Triarch Laboratory making instructional microscope slides. This job fell victim to the Depression, but an interim appointment as a biology*

professor and athletic coach at Greenville College finally let me enter the teaching profession—the profession that has brought me much satisfaction and also enabled me to carry out the research that has made high quality seedless watermelons widely available.

Agnes doesn't have a slogan, but there are things she believes in and finds comfort in—things that have been a large part of who O. J. Eigsti is and what he has done. Although not easily put into words, I remember Agnes saying, "Life has been good. I appreciate every day as it comes. Never a dull moment in all of the years gone by. I am a friend of the Land. I enjoy sunrises and sunsets and all that goes in between. New moons. Full moons. The miracle of growth. Seeds. Humans. The Mystery of Life! A Super Power. 'Let me live in a house by the side of the road / And be a friend of Man.'"

Chapter 2
Why Watermelons?

Why would O. J. Eigsti, or anyone for that matter, be compelled to spend a lifetime trying to improve something as simple and arcane as watermelons? What motivated him? Why did he bounce back time after time after suffering substantial setbacks? How did he manage to regroup after being unaware of or misinterpreting one or more of nature's ways of doing things? How was he able to overcome obstacles in the marketplace? Why didn't year after year of disastrous bottom lines wipe out his dream that seedless watermelons would some day be financially successful?

The answers we seek require more than a passing glance. They lie deep within O. J.—so deep that they may be forever invisible to all who have yet to know the satisfactions that come with growing things, with tending seeds from sowing to harvest, with husbanding livestock from conception to market.

Nor is it easy to organize these pieces of our jigsaw puzzle based on time or geography. In our case, one may believe they came with growing up on 160 acres, a "quarter section," of excellent corn-growing land in north central Illinois. The homestead had no indoor plumbing; there was only a pitcher pump to draw water from a cistern beneath the outside porch. The second floor had no heat, and more than once O. J. woke to find a snow "blanket" covering his bed.

There was a large American chestnut tree near the home that O. J. remembers as much for the prickly burrs that protected the chestnuts as for the nuts themselves. The fate of this tree, a much appreciated part of the early American landscape, adds its own paragraph to this story.

O. J.'s generation was the last to enjoy the taste of the nuts those burrs protected. When the chestnut blight swept across the eastern third of the United States, what were once among the forest's most stately trees were reduced to scattered remnants that survived scarcely long enough to become saplings that produced only a few, if any, chestnuts.

Bioscientists have since been challenged to enable those sorry representatives of a once noble species to again live to maturity. One effort has been going on for more than twenty years at the National Colonial Farm south of Washington, D.C., on the banks of the Potomac River. The project began with subjecting chestnuts to nuclear radiation under conditions that promised to rearrange one or more of the genes that control health and vigor. After they received a prescribed amount of radiation, these chestnuts were planted at an appropriate site on the Farm.

O. J. was a charter member of the organization that created the National Colonial Farm, and he made it a point of not only supporting this project financially, but, even more important, whenever possible he visited this experimental chestnut grove while in the Washington area.

Who knows how many of the things we do later in life, who we are, and what we struggle to achieve, are extensions from our earlier experiences? Perhaps O. J.'s interest in those experimental chestnut trees was conceived in 1920 when Uncle John gave twelve-year-old O. J. and his brother Willis six young walnut trees. Uncle John did not charge them for stabling and feeding their horse during the day while they attended the Maple Grove School. One evening, Uncle John had the six walnut trees ready for them to take home. There was no asking if they wanted them. The trees were planted beside a granary that has long since disappeared, but three of the trees are still there. Even to the point of being sentimental, O. J. hopes that these trees will be preserved as long as possible. He remembers Uncle John as being both a conservationist and a very particular gardener. Those walnut trees had originally been planted by the squirrels—in the wrong place.

O. J. tells us that at heart he is a farmer. His professional work has been tuned to the soil. He grew up accustomed to doing whatever chores were appropriate for his age. He didn't mind keeping the cookstove supplied with wood or the heating stove filled with coal. But he felt differently about milking the cows, particularly in the winter, when they were stabled indoors and their swinging tails were likely to be loaded with manure. All farming was done using horses and they had to be curried and their stalls cleaned. O. J. remembers nothing more stinky than a hog.

He tells us he has shucked corn for weeks and still has those husking pegs. Sometimes the corn was down and had to be picked up before the ears were thrown against the backboard on the wagon. The soil on the

Eigsti farm was a heavy gumbo and if it was worked at the wrong degree of moisture, it was almost impossible to restore it to the proper tilth. Usually, the fields were plowed in the fall so the winter's freezing and thawing would put the soil in better condition for spring planting. The sulky, one-bottom plow pulled by two horses was like the ones the Amish use today, but it was eventually replaced by a two-bottom plow that called for six horses, three in front and three behind.

The two-row corn planter was slow and the planting could be done only when the soil was in just the right condition. O. J. spent many long, hot days walking behind a one-row cultivator pulled by one horse walking between the rows of corn. Planting and cultivating twenty acres was a huge task. When ready to harvest, the wheat and oats were cut, bound into bundles, and dropped in bunches of five or six. Then each bundle had to be stood on end to form shocks with two bundles splayed out and crossed to make a weatherproof roof. Pride in one's work required each shock to look exactly like the rest and to withstand any storm that might come during the weeks when the grain was drying. On threshing day, the bundles were pitched onto wagons and hauled to the threshing machine. Threshing was a community affair enjoyed for its sociability much more by the men. Farm women were bound by tradition to compete for serving the best threshing-day dinner.

But O. J. tells us that life on the family farm wasn't all work and no play:

> *In the summertime, we would ride our bikes to go swimming in Mud Creek. The swimming hole was near a bend in the creek and under a large fallen tree. In those days the water was clear and swimming was a delight. Each year on the Fourth of July we would go to Mackinaw Dells for a picnic – the Litwilers, Yordys, Bachmans, and the Eigstis, Uncles John and Joseph and us. Quite a crowd! We'd spread out the food on paper or cloths laid on the ground. No such thing as picnic tables in those days.*
>
> *I remember the fine sand beach and how clear the water was. Then one year to our surprise the beach was gone. This was the beginning of soil erosion. Before, the land along the creek was left for woods or pasture. Then farmers began cultivating land right down to the creek bank – and there went the clear water. At least I remember how it was before soil erosion began, somewhere around 1914 or '15. Today no one would want to swim in that creek! In the winter we had ice skating. We'd team up, the Yordys, the Litwilers, the Bachmans, and my brothers, and dam Praine Creek to have a place to skate. But some nearby kids didn't join us. Those were good days of fun and games with no fights or trouble with each other. And I remember Aunt Lena giving us a lunch in the middle of the afternoon. It was always a large slice of*

homemade bread that was delicious, but she insisted on spreading a lot of butter before the jelly. To this day I don't like that combination!

The Eigsti farm was not disadvantaged. O. J.'s father was among the first in Illinois to raise purebred Brown Swiss cattle. He held out for this breed because it combined the production of both beef and dairy products. To him, the Brown Swiss were superior to Shorthorns that were just for beef and Jerseys that were just for milk. O. J. is more inclined to remember that the cows had to be milked by hand every day, morning and night. And someone had to turn the separator by hand to get the cream for butter.

Two scenarios illustrate O. J.'s close ties with the evolution of agriculture throughout most of the twentieth century. Both settings are on late summer days, the first in 1920, the second in 1985; the first on the Eigsti family farm, the second on the farm of Duane and Laurie Johnson in Knox City, Texas. The first occasion opens very early in the morning. O. J. is helping to pack several dozen dressed squabs in ice. Yesterday these nearly adult pigeons were taken from their nests in specially designed quarters above the chicken house. After crushing their heads as humanely as possible, all of the feathers, particularly the pin feathers, were plucked, one by one. With no further "dressing," the squabs were ready to be packed in ice and delivered to the railroad station to be rushed to market in Chicago.

This was only the first mission of the day. Each bit of space in the family car was being packed with produce from the farm. All fresh vegetables from the garden, tender green beans, red beets washed and tied in bundles, crisp cucumbers, summer squash, vine-ripe tomatoes to be eaten skin and all. Perhaps the first apples of the season were ready to harvest or other fruit from orchard or "truck patch," and eggs, butter, a few dressed chickens, ducks, an occasional goose or turkey specially ordered by a housewife or butcher along the streets of Peoria, where the Eigstis peddled their farm produce.

O. J.'s father did not always head up this operation. Sometimes his older brothers took charge. But the money collected along the way was almost sure to end up in the bank in which O. J.'s father had a special interest, not only in the Eigsti account, but in the bank's bottom line. This was very much a family enterprise. All members took part, not as "hired hands," but as partners in a common interest. Each did what needed to be done. This was the expected way of life, a way of life that differed little from that which O. J.'s ancestors brought with them from the "Old Country." O. J. gives us glimpses of these trips to town:

I remember how we would dress chickens, pigeons, squabs, turkeys, geese and ducks and take them to Peoria to several places. One was the Creve Couer Club. We brought home hard rolls and bread that was discarded from the tables to feed the chickens and pigeons. On the way we had to go through East Peoria down a very steep hill. It wasn't paved and coming home we couldn't make it in high gear.

On the way to Peoria we passed the Holt Manufacturing Company, which later became Caterpillar. The company came from southern California, where the crawler type of machine was developed to cultivate the sandy land of the desert. During World War I, the Holt grounds were used for training soldiers and I think they also built the first tanks. Peoria was also famous for the production of whiskey. I understood that my grandfather took wagon loads of corn to the brewery and got gold in payment, and in those days it was OK to bring home cases of beer for the thresher crews.

The second scenario opens a full seventy years later. Here at the Johnson ranch at Knox City, Texas, we find O. J. visiting with each of the drivers of some dozen or so eighteen-wheel semis loaded with seedless watermelons awaiting their final clearance. The trucks were going to Boston, Vancouver, Winnipeg, Los Angeles, and Florida. Some of the melons were taken to Los Angeles, put on a boat, and shipped to Japan. O. J.'s direct involvement here was that the seed for these melons came from the American Seedless Seed Corporation. Common to both of these scenarios is the motivation to produce something people need.

These scenarios, set as they are at the opening and closing of a century encompass a large slice of how people have earned their living as farmers throughout the twentieth century. Thus, contemplating who O. J. is and what he has done provides a window of opportunity to see big things through the perspective of one person. Biggest of all is the increased complexity and the regulation of the channels that connect the producers of foodstuffs and the consumers. And with each change, new people are enabled to ride piggyback, to add their services to the process, and to be paid accordingly. Initially, only the integrity of the producer protected the consumer. Discrepancies or shortcomings were affairs between the people who produced and those who consumed. Short-haul delivery was a part of production; profits went directly to the producer. There was no sales tax, no income tax, no fees to be paid for a business license or permit.

Customers took for granted the quality of the Eigsti family's farm produce. In turn, meat and dressed poultry, fruit, and garden vegetables they offered their customers were treated with the same care as those that ended up on the Eigsti dinner table. The ethics and traditions of the Mennonite faith permitted nothing less.

So, why not give watermelons top billing? The watermelon has a long and happy history. It is believed to have originated in tropical Africa. Its early ancestors were inedible until mutations changed bitter substances in the fruit. The missionary David Livingstone described edible watermelons in his report of an expedition to the Kalahari Desert. The Hebrews pined for watermelons in the wilderness after becoming accustomed to them in Egypt. The early Spaniards brought melons with them to the New World and they were much welcomed in the gardens of colonial America.

Watermelons are so sweet and juicy that we think of them as a fruit and serve them as dessert or picnic fare. Their actual botanical relatives are cucumbers, pumpkins, and squash, including, of course, the ubiquitous zucchini. From a genetics point of view, the watermelon is a veritable gold mine. The genome of the species provides a wonderful shopping list of inheritable traits. There are genes that call for melons that weigh more than a hundred pounds, or only a few pounds if they are to be designed to fit into home refrigerators. They can be round, oval, elliptical, or oblong. There are genes for different shades of pale green to almost black and for solid, striped, or marbled skin pattern. Scarlet is the favorite color of watermelon flesh, but there are also genes for red, pale red, yellow, cream, and white. The seeds may be black, brown, or white, sculptured or plain, engraved or with spots and colored edges.

Seed catalogs list only a dozen or so of the hundreds of varieties of watermelons that have been developed. Variations from the "standard" varieties seem to reflect the amazing creativity of the people who write those mouth-watering passages for seed catalogs. In actual practice, the same variety of melon may appear under names that reflect differences in writing skills more than characteristics of the melons themselves. The name of one well-established variety traces to the rind pattern resembling that of a rattlesnake's skin.

The wide variety of genes in the watermelon species has both advantages and disadvantages. Sorting out the set of genes that will yield the best variety of melon is something like dealing a bridge hand. Just as there must be exactly thirteen cards in each hand, each variety of watermelon requires a set number of chromosomes with an equally prescribed number of genes on each chromosome. The trouble is that any attempt to "stack the deck" to give one variety of watermelon a preferred gene is sure to affect other characteristics. Furthermore, a wide array of genes affects a watermelon's resistance to disease and the ability to cope with soil and environmental conditions. An attempt to breed a preferred gene into a variety of melon includes the risk of introducing a gene that will bring a totally undesirable trait. Actually, managing to bring together in one variety of melon the most advantageous collection of genes became the primary reason why O. J. Eigsti was able to produce a seedless watermelon that competes favorably with all other commercial varieties. The many difficulties that were encountered along the way explain much of why it took so long.

There were many obstacles. Twenty-five years ago at a meeting of the National Watermelon Growers Association, the members let O. J. know that there was very little interest in seedless melons; in fact, the general feeling was that they were neither needed nor wanted. This attitude reminded O. J. of the early reactions of corn growers to the introduction of hybrid corn. In the '20s and '30s, no one needed or wanted hybrid corn. Before the '30s were over, however, everyone wanted hybrid corn; by 1940 the change from open pollination to hybrid was complete. Today, O. J. is sure the seedless watermelon will repeat the hybrid corn story.

Actually, the development and ultimate commercialization of hybrid corn created a wealth of knowledge and know-how that permeated O. J.'s mind in countless ways. He grew up with the hybrid corn industry, shared its adolescence, and was closely involved with its blossoming into a very financially successful enterprise. He read the related literature and was on his way to devoting his life to this industry in 1950 when he signed on as a geneticist with the Funk Brothers Seed Company. The decision to shift his interests—to bet his future on watermelons rather than corn—is the reason this story is being told.

The development of seedless watermelons shares many similarities with the development of hybrid corn. Both stories beautifully illustrate how science-related discoveries and inventions occur. Things begin when someone recognizes an event, circumstance, or state of affairs in nature that is contrary to the expected. This happens, of course, most often to people who are very familiar with a particular body of knowledge related to the unexpected event.

For hybrid corn, the surprising vigor of offspring born from crossing two different strains of inbred parent plants was the initial unexpected event. This hybrid vigor phenomenon, also known as heterosis, was mentioned, or at least hinted at, by such well-known mid-nineteenth-century scientists as Charles Darwin and Gregor Mendel. The "to be expected" point of view was that offspring born to closely related parents were likely to be weak, deformed, and much less vigorous than their parents.

In the seedless watermelon story, the appearance in nature of occasional plants (or animals) with two or more times the usual number of chromosomes, polyploidy, was the surprising, the unexpected event or circumstance. Eigsti tells us in his book, *Colchicine*, that recognition of this phenomenon also goes back to the nineteenth century.

The next phase of discovery and invention seems to shape up when an observer becomes curious, mentally uncomfortable, and anxious to explain the troublesome, nonconforming circumstance and bring it into the realm of the expected. This is usually accomplished by proving that the disturbing event is nothing more than heretofore unrecognized interactions among familiar scientific concepts or principles. Of course, there are those ultra–high level, Nobel Prize–winning situations that lead to the identification of a new concept or principle.

For both hybrid corn and seedless watermelons, although much precision has been added to descriptions of the underlying phenomena, many have not yet been accounted for with complete satisfaction. For seedless watermelons, however, the discovery that colchicine can cause polyploidy is vitally important.

Using colchicine-induced polyploidy to produce seedless watermelons probably happened first in China or Japan. William James Beal, following up what he called the buoyancy of hybridization, "probably made the first controlled crosses between varieties of maize for the sole purpose of increasing yields through hybrid vigor," according to *The Encyclopedia of Agricultural Science.* "Beal's achievement was without knowledge of the genetic principle involved," and it may be that the idea came from having read about work done in 1876 by his friend Charles Darwin.

Improving the precision of the description of a principle, together with using the principle to solve technological problems, marks the closing phases of science-related discoveries and inventions. The work of O. J. Eigsti and his many seedless watermelon compatriots illustrates this phase of science-related discovery and invention.

The huge amount of human resources that have been invested in developing the hybrid corn industry makes it a much longer story. For example, one of Beal's students was Perry G. Holden, who went on to develop a large seed corn business in Illinois. He believed firmly in selecting the very best ears from the current year's crop to be seed for the next year. For some reason, Henry A. Wallace (later to become Franklin D. Roosevelt's vice president) questioned the soundness of this practice when he was only sixteen years old. To prove his point, he saved thirty-three of the very best ears from Holden's best variety of corn. He then planted half the grains on each ear under conditions that would enable him to keep track of the separate yields from each ear. The results showed wide variations in yields, from thirty-three to seventy-nine bushels per acre. This strengthened Wallace's belief that something more than the visible characteristics of a seed influenced the productivity of the next generation.

The year was 1904 and other curious agriculturists were following up Beal's description of the buoyancy of hybridization. Two investigators, Dr. George Shull at the Carnegie Institution for Experimental Evolution on Long Island and Dr. Edward East at the Connecticut Experiment Station, tested a particularly interesting hypothesis: If the offspring of different inbred parents were unusually vigorous, then the more the parents differed genetically, the more vigorous the offspring would be. To test this hypothesis, two varieties of corn were inbred for several generations in order to enhance the differences between the two varieties. This called for some courage in the face of ridicule because it was generally believed that inbreeding would weaken both of the ultimate parents.

This seemed to be the case, but when the two inbred parents were cross-pollinated to produce a new generation, the results were amazing. As expected, many of the next generation of corn plants were freakish, but there was an impressive increase in the total yield of marketable corn. There could be no further doubt but that hybrid vigor was real.

This episode in the history of hybrid corn parallels Kihara's proof in the 1940s that it was possible to use polyploidy to produce seedless watermelons. There are additional similarities as well. In both cases, it took highly dedicated believers to work out the ways and means required to transform fundamental knowledge into commercial success.

There is something more that makes watermelons unique. Among all of the fruits and vegetables that farmers produce, there is something about watermelons that has long made them strangely susceptible to being stolen. For some reason, people who would never think of stealing anything else consider watermelons fair game. The many stories that stem from this state of affairs are difficult to trace to their true origin. One example, however, is interesting because it tells us other things. As remembered by Irene Browning, many years ago a new schoolmarm in rural Colorado, it all began when two male colleagues "in whispered tones, invited me to go stealing watermelons. They also invited Mildred, who was also new that year. If they believed it was acceptable, I accepted the invitation thinking it might be fun."

Rather elaborate plans included stopping by the home of another teacher, who "gave us a royal send off and told us to be sure to bring him a watermelon. We drove out of town where Dean knew there was a good watermelon patch located well off the main road. He turned the car facing out in case we had to get out in a hurry. We then walked along a cornfield peering into the darkness trying to find the watermelon patch. I happened to look over my shoulder and saw a flashlight coming our way. I just gasped and pointed. Dean said, 'we gotta get out of here.' The light fastened on us and a piercing shot rang out. From then on I don't remember much, but it was a matter of life and death.

"I was terribly frightened going to bed that night. The world in general just didn't seem safe. But the morning light was reassuring, and even though sworn to secrecy, I had to tell someone. I poured forth my tale to Mrs. Galloway, my landlady. Was I ever chagrined! Instead of being sympathetic, she just laughed. 'That was a frame-up, Irene. Those two fellows do that to the new teachers every year. No one around here would shoot and there are no watermelon patches.'"

Why watermelons? Again, why not? A life needs a focus, something to look forward to. For O. J. Eigsti, the success of the seedless watermelon has met that need very well. It allowed him to be scientist, farmer, and educator. It has taken him along the world's highways and brought breadth rather than narrowness, hope rather than resignation. Of course, frustrations have competed with satisfactions, rejections with acceptances. Wise decisions have been sacrificed to cover the losses from bad; synergetic gains from random coincidences have been wiped out by circumstances that could not be avoided. But such is life—and the life of O. J. is no exception.

A final answer to the question, "why watermelons?" comes from Judith C. Glordan's article, "Science in the 21st Century: Harmony for the Millennium," which appeared in *Chemical and Engineering News* for December 2, 1996. O. J.'s work on the seedless watermelon resonates perfectly with "one of the greatest challenges to face the scientific community at the end of the twentieth century [and] is no different than the challenge at the end of the 19th: overcoming the prejudices and fears encompassed by the classic debate of which is better and more needed—fundamental or applied research . . . and which is more important and should gain greater recognition and respect—academic or industrial research?"

Ms. Glordan's audience is primarily the members of the American Chemical Society. How her topic is presented to the public, however, is better documented by a segment of a widely viewed magazine-type television show broadcast on Febraury 9, 1997. This show featured the amazing success of a comedian and his writers. Examples included momentary reference to the seedless watermelon as proof that scientists pursue ridiculously useless goals. This was followed almost immediately by the comedian creating humor out of young people dealing with a sexual taboo. Both bits were funny, supposedly, but made less so when one weighs the goals being pursued by the comedian and his sponsors against those of O. J. Eigsti and his colleagues.

Neither the comedian's influence on his audience's attitude toward science nor the psychosomatic outcomes of tampering with sexual taboos is at issue here. What is at issue is how one's life reflects personal values. In spite of the ridiculously wide gap in their financial rewards, the comedian's success does not leave O. J. regretting his choice of a lifetime goal, nor does it take anything away from our answers to "why watermelons?"

Chapter 3
Colchicine: A Fantastic Drug

A NEWSPAPER CLIPPING that is almost too yellowed to be legible has a photo with the caption: "Dr. Eigsti of the University of Oklahoma with a sweet potato plant which has been treated with colchicine." The title of the article is "Plant Chemistry for the Amateur," and the contents tell us about O. J. wanting to share his research interests with everyone. The clipping continues, "chemical gardening has not been taken up generally by the average, serious-minded home gardener seeking practical results and good, fresh vegetables for the table. Nevertheless, many amateurs have been participating with Dr. O. J. Eigsti . . . in an experiment with a drug called colchicine . . . which when applied to the growing parts of plants causes changes within the cells. These changes form the basis for new types of plants, called mutations, which are deviations from the parental type."

This clipping tells us that O. J. encourages curious people to obtain a sample of colchicine suspended in a salve-like mixture and see "what would happen if you treat a squash, a cosmos, a marigold or some other plant." At the same time, he alerts amateurs to the properties of colchicine and warns them to avoid actually touching the colchicine preparation. He also warns against building up high hopes by saying, "a botany department survey of results achieved by about 100 amateur botanists shows that about one of every five persons experimenting with colchicine is successful in bringing about some type of change. Of course, there are disappointments ahead in some cases." We can add that O. J.'s experience as a research scientist has required him to keep his hopes up, to hang in there year after year when his success record has been far lower than one in five.

Why give such special attention to colchicine, a substance that means little or nothing to nearly everyone? Attempting to tell the life story of O. J. Eigsti and not refer repeatedly to this derivative from the autumn crocus would be equivalent to telling the life story of a cocaine addict and not introduce cocaine, a derivative from the coca plant. Obviously, I don't mean to say that Eigsti is dependent on colchicine for psychosomatic reasons, but this substance has played a leading role in the research that has made seedless watermelons possible. Colchicine enables scientists to get at the very center of cell division, the process that is essential to the development and growth of all living tissues.

Memories of school biology courses almost invariably include microscopic examination of samples of plant and animal tissues that have been sliced very thin and treated with special dyes to make visible the cellular components that are involved in cell division, the process called mitosis. For the special case where egg and sperm cells reproduce, the term changes to meiosis. In both cases, the most readily observed "actors" in the cell division process are the chromosomes, the oddly shaped objects that carry the genes, the molecules that have the fantastic property of handing down from generation to generation the "blueprints" for the structures and processes that add up to a new individual.

But there was something different about mitosis and meiosis as these processes were taught while O. J. was growing up. At that time, innovation added excitement as people in research competed to be first to describe accurately what chromosomes and genes really are, what they can do, and how mitosis and meiosis really occur.

This may explain why O. J. focused on one feature of the cell division process when he began the research that was a part of his doctoral work at the University of Illinois. The feature is interesting from a rather special point of view. It is to be expected that if each parent's cells contain, say, eleven chromosomes (as is the case with watermelons), then it is to be expected that when a cell divides during the growth process, each "daughter" cell would also contain eleven chromosomes. And this is what happens during normal mitosis in watermelon tissue.

But there can be exceptions. Things happen in nature that result in "daughter" cells receiving two or even more complete sets of chromosomes, a phenomenon known as polyploidy. Furthermore, this state of affairs is handed down during each additional cell division, and continues from generation to generation. For example, there can be individual plants in which the cells of all tissues contain four complete sets of chromosomes. This condition is called tetraploidy and the individual plants are known as tetraploids.

Here is where colchicine enters the O. J. Eigsti story and explains why a solution containing colchicine was dropped on those watermelon seedlings in that campus greenhouse. Under certain conditions, colchicine causes the cells in the growing tips of watermelon seedlings to become tetraploid. The daughter cells resulting from further cell divisions will contain forty-four chromosomes, or twice the usual number. Due to the reduction phase of meiosis, the egg and sperm cells produced by a tetraploid plant will each contain twenty-two rather than the usual eleven chromosomes.

How this special state of affairs makes possible the seedless watermelons that we now see displayed on supermarket produce counters is the story I will tell. For now, to emphasize O. J.'s "addiction" to colchicine, I cite only the hundreds of hours he and a coauthor spent tracking down the 1,652 citations that made possible their book, *Colchicine—In Agriculture, Medicine, Biology, and Chemistry*. This work brought to O. J.'s mind just about everything that was known about this fantastic drug and what it can do. Indeed, writing the book and getting it published was a major episode in O. J.'s life and deserves the emphasis it receives in this volume.

It is particularly thought provoking to find a scientist in the twentieth century using a substance for research purposes that has such a very long history. The earliest medical texts report use of colchicum, an impure form of colchicine, in treating gout, rheumatism, and "tumors" that had not yet "spread into the body." The name comes from its primary source, Colchicum autumnale L., commonly called autumn crocus. Another common name, "naked ladies," is prompted by the appearance of flowers out of the ground without leaves. Thirty-six different species flower in the months of September to November. The plants are particularly naturally abundant in the ancient land of Colchis at the eastern tip of the Black Sea. They also appear in cultivated flower beds wherever environmental conditions are favorable.

The active substance, colchicine, can be extracted from buds, leaves, flowers, stems, and the underground corm. Although it is an amazingly active substance, only very small quantities of the pure crystalline colchicine can be obtained from large quantities of crude plant material; usually less than 0.01 percent and rarely more than 0.2 percent.

It is difficult to talk about a substance that has healing properties—particularly in the context of the seedless watermelon story, in which colchicine-induced polyploidy plays the key role—without commenting on the chemical mysteries of the life process. The chemistry within the replication of chromosomes during mitosis, in itself, boggles the mind. In effect, we know now that each chromosome is a collection of separate genes, with each gene being a unique kind of "cellular molecule." During cell division, two new molecules cannot appear where there was formerly only one without each being the product of its own unique chemical reaction. A chemical reaction involves a source of raw materials coming together under unique reactant conditions; the taking apart of highly specific molecules and reassembling the fragments in an equally uniquely prescribed architecture.

The logistics are awesome. One can only wonder how the required supplies of the essential reactants, the raw materials, are brought to the scene of action. And how the reactant conditions that permit one kind of molecule to be replicated are kept from interfering with other similarly unique reactions within such a crowded environment. As every chemist learns, usually the hard way, allow even the tiniest trace of a contaminant or the slightest deviation from the prescribed reactant conditions, and the yield of desired product is sure to be messed up. And yet within the unbelievably intricate cellular chemistry, the presence of colchicine molecules brings about a predictable effect.

The colchicine molecule is a specific bonding of atoms—twenty-two carbon, twenty-six hydrogen, six oxygen, and one nitrogen. A fundamental premise in chemistry is that the properties of a substance derive from the kind, number, and arrangement of the atoms that make up its unit structures, usually molecules. The greatest of all challenges faced by anyone who sets out to understand, manage, or manipulate the chemistry of any or all of life's myriad processes, is to enter the domain of atoms and molecules, in effect, to visualize and think like each kind of atom in order to sense why it is attracted to one and repelled by another while the atoms are being reshuffled from molecule to molecule.

The challenge, in itself, is unsettling. Our earliest chemists found refuge in invoking mystic or magic forces. The medicines they administered were often seen as paying tribute to evil spirits or sorcerers and sorceresses who held power over sickness and health, life and death. When Eigsti was treating those watermelon seedlings in that campus greenhouse with colchicine, he was millennia beyond mysticism and magic; but those of us who know him well can say he has never isolated himself from the mystery and majesty of chemistry's role in what it means to be alive.

The book on colchicine, which O. J. is so proud to have written, can be seen as one of the thousands of efforts to bring to our minds features of the infinitely magnificent drama that is the mystery of life. The metaphor is wonderfully appropriate and I use it here to help us appreciate what it means to O. J. to be able to match his wits with one of life's most fundamental processes. The book suggests attending a theatrical production, a ballet, perhaps, in which colchicine plays the leading role.

The program we receive as we enter the theater gives a brief and, as usual, an almost meaningless distillate of what we will see. We read, "The number of chromosomes in plant cells is frequently doubled after treatment with colchicine." We aren't told very much about what chromosomes are and even less about what kind of weapon or tool colchicine is. But we know enough about the author to understand why the rising curtain reveals a setting in the middle of a watermelon patch, and that life as it is lived in a watermelon plant is sure to be involved in the action on the stage.

The opening scene has several thousand dancers milling about randomly. They fill the stage completely. Each has a unique costume, misty, almost cloud-like, and we notice there are eleven different colors of them. In time, for no apparent reason, the dancers form eleven groups and we notice that all of the members of each group have the same color costume. As the groups form, they line up in a definite order near the center of the stage where, almost miraculously, each group doubles itself. Now we see each of the two groups of eleven dancers work their way to opposite sides of the stage. Here the groups disassemble, break up, and as the curtain comes down on the first act we see two groups of the same several thousand dancers milling about on opposite sides of the stage in very much the same fashion as at the opening of this production.

When the curtain rises for the second act, everything continues on the stage exactly as it did for act one. But only up to the point when the dancers are lining up at the center of the stage. At this point the stage is flooded with new actors. All of them wear the same costume and our program tells us their name is colchicine. At this time, chaos reigns on the stage. There is much "body contact." When things settle down, we see there are four groups each containing eleven similarly costumed actors where there had been only two groups. Now, the actors form two sets, each containing eleven groups, with all of the dancers in each group wearing the same color costume. The groups now assemble at opposite ends of the stage. At this point, the action is the same as at the end of the first act, except there are twice as many groups of actors on the stage.

Yes, the dancers represent genes, and the groups they form, chromosomes. The dance they perform is mitosis, first without and later with the action of colchicine. But the metaphor is far from adequate. There is nothing said about the "director" in the wings who instructs each of the "genes" and "chromosomes," who tells them what to do and when to go where so that the show will go on. And there is something that suggests deception to see four "chromosomes" appear out of thin air where there had been only two. After all, we know that each "dancer" is put together from dozens, perhaps hundreds or even thousands of separate atoms that have to come from somewhere. Nature is even more unrelenting than a dance director in insisting that each atom must occupy its own unique position if the resulting molecule is to have the properties we expect it to have.

And the show will go on. Colchicine has caught and held the attention of scientists for thousands of years. And there are countless other substances that play equally vital roles in the chemistry of life. All of which is to say that this phase of the O. J. Eigsti life story does not end with the twentieth century, but describes research as it will continue in the new millennium.

Chapter 4

Genetics is for Everyone

O. J. Eigsti is a scientist, but with a special interest in genetics. We are all geneticists in a way. We are all curious about how our parents and grandparents have influenced how we look and the kind of person we are. Genetics is pretty much a twentieth-century term, but this is a very old branch of science. Ideas about how characteristics are handed down from generation to generation go back beyond Biblical times. In fact, some of these ideas continue to influence the beliefs of many people. We cannot tell the story of O. J. Eigsti and the seedless watermelon without a review of how the current knowledge of genetics came to be.

"Like begets like" has been a universally accepted concept throughout human history. There are equally long-standing "amendments" to cover cases where offspring are not like their parents. One classic example that has been handed down from biblical times is also an example of applied genetics. It was a case where the "geneticist" would benefit if he could make a new generation of sheep and goats to include a higher proportion with "speckled and spotted" coat colors. He accomplished this by surrounding the watering places during the breeding season with "fresh rods of poplar and almond and plane" that had been worked over to reveal threatening patterns of speckles and stripes.

This very early "geneticist" also arranged to have only the stronger members of the flock be frightened by the strange rods. And it may have been more than in a dream when he saw that "all the goats that leap upon the flock are striped, spotted, and mottled." In any case, we know that he inherited most of his father's livestock. To this day, many people believe that their offspring's inheritable characteristics can be influenced by environmental circumstances that accompany their conception and birth.

When O. J. Eigsti was a farm boy, it was common practice to save the largest potatoes for next year's "seed" and to make do with the smaller potatoes at the dining table. And only the biggest, best formed ears of corn were to be saved for next year's planting. Furthermore, O. J. remembers removing by hand all of the round or misshapen grains at the ends before dropping the ears into the mechanical sheller. And much has been made of nature's ways to ensure that only the strongest, the most noble male is sought after by all of the females during the rutting season.

This state of affairs beautifully illustrates the public's attitudes involving not only genetics but all branches of science that prevailed during the start of the twentieth century. Stated bluntly, new science-related beliefs—new ways to do things where science was involved—met stiff resistance. It wasn't easy to entertain the notion that causes other than heredity might be responsible for some potatoes or ears of corn being larger than the others in the same field, or one buck being the strongest and most aggressive deer in the forest.

There is another way the "like begets like" concept can be misleading. Contrary to the spectacular differences we see in the many breeds of dogs today, there was a time when all of the dogs in the world looked alike. We can say the same for horses, chickens, pigeons, rabbits, cats, and all other domesticated animals. It is possible to bring about amazing differences in the breeds and varieties of domesticated animals and plants. For example, perhaps someone who raises cats notices that a few kittens in one litter have some interesting characteristic more highly developed than their siblings. These kittens are selected to be the parents of the next generation. Among their kittens, only those that excel in the desired characteristic are selected to be the parents for the next generation. Continue this selection strategy through many generations and the final offspring would differ greatly from the original generation, especially if the breeder "selected" for combinations of several characteristics.

This selection strategy can be very effective. It is said that one cattle breeder became so thoroughly smitten by his idea of what a perfect cow should look like that he painted a large picture of an "ideal" cow on the side of a barn. His neighbors were puzzled because they had never seen a cow like that. But, sure enough, years later the painting did not differ very much from the actual cows in the farmer's field. As strange as it may seem, toy poodles and Saint Bernards do share a common ancestor.

Some new breeds or varieties of plants and animals trace their origins to differences between parents and their offspring that appear in only one generation. A red flower can appear when all other flowers on the same stem are white. A black squirrel can show up in a city park where all other squirrels are gray. Furthermore, these totally new characteristics reappear in the offspring of these sports, a term used to describe individuals that have a characteristic differing markedly from that of their parent that can be handed down to future generations.

All of which is to say that while O. J. Eigsti was growing up, people had their beliefs about why offspring usually, but not always, resemble their parents. These beliefs enabled them to manage, and to a certain degree control, the handing down of characteristics from generation to generation. Everywhere throughout the world, farmers, gardeners, herdsmen, and pet fanciers have taken great pride in whatever breed or variety best met their specifications. All of these spectacular achievements were based almost entirely on the selection strategy; all relied on the notion that "like begets like," supplemented by the mysterious and abrupt appearance of offspring that differed markedly from their parents.

At about the time O. J. was born, many of these beliefs were being questioned. The cause was the reappearance of ideas first published in the middle of the 1800s by Gregor Mendel. Before retelling the story of this Austrian monk's work with the genetics of garden peas, however, I cite the less common story of Maude Slye's investigations of the genetics of mice early in the 1900s. Her work could well have had an equal if not even greater impact on the development of this branch of science.

I begin this part of my story of O. J. as a geneticist in reverse chronology. It is the summer of 1996, forty-seven years after we met O. J. Eigsti coming into the university greenhouse carrying trays of watermelon seedlings. Today he has invited us to join him in his daily, almost ritualistic walking tour of his laboratory. O. J. wears no white uniform; in fact, he would promptly throw it away were he offered one. He carries no microscope or other high-tech apparatus, no computer, and no notebook to collect records required by a government grant or industrial sponsor.

Our first stop is by the side of a test plot where some 100 watermelon plants are just beginning to show their first flowers. There are no weeds and the soil is perfectly cultivated thanks to the Amish caretaker. With the family dining table in mind, missing watermelon plants have been replaced by occasional tomatoes, ground cherries, cantaloupes, or sweet peppers. The important feature is that watermelons have been grown each year for seventeen years in this test plot. By now, the soil is sure to be infested with every disease that threatens watermelons and thereby will automatically "select out" those that are not disease resistant. The seeds that produced these plants came from the survivors of twenty previous generations. Later this fall, one melon will be cut from each plant to see if it fulfills O. J.'s hopes: flesh with bright crimson color, sweet and tasty, a shape that doesn't crush easily, and rind tough enough to head off shipping problems. The melons that come closest to fulfilling these hopes will provide the seed for next year's planting.

We are not truly alone on this walking tour. The spirits of Gregor Mendel, Luther Burbank, and the generations of geneticists and plant and animal breeders who have struggled to improve humanity's sources of food and fiber are around us. In a very real sense, their lives continue in O. J.'s mind. No longer are they nothing more than footnotes in our publications, citations in university lectures, or anecdotes in school science lessons. Their lives and what they struggled to achieve are now guides to where O. J. wants to go.

We move on to a larger field. Here are 500 more tetraploid watermelon plants. This is new ground with far less chance of being contaminated with disease or suffering from depletion of essential trace elements. There is greater opportunity for the full genome of the watermelon species to be revealed. Again, one melon from each plant will be "sacrificed" and seeds from those that show the most promise will be kept for next year's planting. The look on O. J.'s face reflects contagious enthusiasm and a boundless faith that what he has spent so much of his life looking for will appear when one of those melons cracks open.

All of this gains credence when we come to the final stop, a makeshift library. There are rows of bookshelves in a room set apart in a one-time turkey shed—a building as long as a football field and filled with things both usable and unusable but too good to throw away. Here are books with names and dates that tell us they were sister Minnie's school science textbooks going back to the early 1900s. Sharing space are university textbooks with such titles as *Plant Physiology*, *Methods of Plant Breeding*, and *Evolution and Genetics*. Their publication dates are in the 1930s and '40s and are intermingled with genetics textbooks that O. J. used in the courses he taught at Oklahoma, Northwestern, and Chicago State universities.

One badly discolored and terribly musty volume brought to mind an uneasiness that nearly surfaced while O. J. touched on both genetic and environmental factors in telling about his hopes and ambitions during the morning tour. The title: *The Situation in Biological Science: Proceedings of the Lenin Academy of Agricultural Sciences of the USSR Session: July 31–August 7, 1948*. The opening chapter is "Address Delivered by Academician T. D. Lysenko on the Situation in Biological Science." The fly-leaf tells us that this book was "Presented to Dr. Eigsti in 1953 by Mohammad Saleem in memory of the discussion with Mr. Faradi." Actually, this "discussion" was a debate on the merits of Lysenkoism that took place in 1953 at the Indian Science Congress held at Lucknow. O. J. defended the negative point of view.

The 631 pages of this book document the Lysenko affair, which as marked by, quoting Lysenko, "A sharp controversy, which has divided biologists into two irreconcilable camps has flared up over the old question: *can characters and properties acquired by vegetable and animal organisms in the course of their life be inherited?*" (italics in the original). Without doubt, O. J. has had to deal with this controversy because it influenced the beliefs and attitudes of his contemporaries both in America and elsewhere. To review specific circumstances would risk misinterpretations and misunderstandings. But I would be remiss if I failed to identify the Lysenko affair as a part of what it has meant to be a geneticist in the twentieth century.

Recalling Lysenkoism leads to thought-provoking comparisons. In some respects what Lysenko preached in Russia during the middle of the century, Luther Burbank professed earlier in the century in America. Burbank believed that "characters and properties acquired by vegetable and animal organisms in the course of their life" can be inherited. (Burbank's beliefs and scientific achievements are discussed in more detail in chapter 5.) But in America, unlike in Russia, nonbelieving scientists weren't sent to Siberia.

Nor were textbooks burned if they failed to point out the errors of which Darwin was guilty. It wasn't that Lysenkoism didn't accept Darwin as a "great naturalist, the founder of scientific biology, whose activity marks an epoch in science." Darwin's sins were political. In the party line preached by Lysenko, "there is absolutely no justification for accepting the erroneous aspects of the Darwin's theory, those based on Malthus' theory of over-population with the inference of a struggle presumably going on within species." It wasn't that Darwin represented something bad, but, "only by further developing Darwinism and raising it to new heights would biological science become capable of helping the tiller of the soil to obtain two ears of corn where there was formerly only one." In fact, Lysenko found much more fault with the "teachings of Weismann, Mendel, and Morgan, the founders of modern reactionary genetics," than with Darwin's attempt to account for the origin of species.

Much of what Burbank did in America was claimed to have been equaled if not exceeded in the Soviet Union. The most prominent name is that of Michurin, whose research led to the term Michurinism and was cited by Lysenko as providing "the foundations of Soviet agrobiological science." When the political implications are considered, however, the term Lysenkoism looms larger than Michurinism. Luther Burbank's beliefs may, at times, have contradicted those of his contemporaries, but the likes of Lysenkoism have never been visited upon America's people.

After more than 600 pages of speeches proclaiming Lysenkoism and defaming Mendelism, Lysenko "lets it all hang out" in the final chapter of the aforementioned proceedings.

> I call upon all Academicians, scientific workers, agronomists, and animal breeders to bend all their efforts and work in close unity with the foremost men and women in socialist farming. . . . Progressive biological science owes it to the geniuses of mankind, Lenin and Stalin, that the teaching of I. V. Michurin has been added to the treasure house of our knowledge, has become part of the gold fund of our science. [Applause.] Long live the Michurin teaching, which shows how to transform living nature for the benefit of the Soviet people! [Applause.] Long live the Party of Lenin and Stalin, which discovered Michurin for the world [applause] and created all the conditions for the progress of advanced materialist biology in our country. [Applause.] Glory to the great friend and protagonist of science, our leader and teacher, Comrade Stalin! [All rise. Prolonged applause]

It isn't easy to sort out how the Lysenko affair affected O. J. and his work on the seedless watermelon. O. J.'s marginal notes and underscored portions of the book are clues. Where the action of colchicine is discussed, for example, we find:

> We insist that such action, which penetrates into the organism not in the course of its development, not through the process of assimilation and dissimilation, can only rarely and only fortuitously lead to results useful for agriculture. It is not the road of systematic selection, not the road of progressive science. The protracted and numerous efforts made in the Soviet Union to produce polyploid plants with the aid of colchicine and similar potent factors have in no way led to the results so widely advertised.

We can imagine O. J.'s feelings while he was underscoring "all these polyploids—wheat, millet, buckwheat—have so far, as a rule, been found to be of small fertility, and their authors themselves have refrained from commending their cultivation for practical ends. . . . Is it not obvious that breeding must be conducted, not with a view to the number of chromosomes, not with a view to polyploidy, but with a view to inducing good qualities and properties?"

There is another way the Lysenko affair may have affected O. J. and in fact, all of us. This affair, with all of its ins and outs, leaves no doubt but that people's science-related beliefs can be used by selfish interests. This is not an example of "political correctness." There is no effort to disassociate political agendas from people's beliefs about nature's ways and means. On the contrary, nature is blatantly interpreted to best serve political interests. In fact, if there is any good to be salvaged from the Lysenko affair, it may come from teaching us to look at proposed beliefs about how things happen in nature from the point of view of what agendas and ill-advised motives these beliefs are being forced to serve.

Viewed in retrospect, the Lysenko affair was a blatantly clumsy, clenched-fist attempt to use a psychological tool to influence people's beliefs and behavior. Influencing people's beliefs about nature, about our place in the natural world, is a strategy far more effective than less subtle governmental decrees or the most carefully crafted political sound bites or the noisiest of public demonstrations. If O. J. and the generation of geneticists he represents were not put to a disadvantage by the Lysenko affair, the reason may well be only the lack of finesse. The potential effectiveness of the underlying strategy remains intact. And for certain, only an "explanation" of life itself may exceed notions regarding the handing down of characters from parents to their offspring as an effective way to precondition people's beliefs and behaviors.

Returning now to a more conventional chronology, I will account for what was in O. J.'s mind not only during our morning tour, but throughout other genetics-related decisions of his career. To set the stage, I begin with the role of Gregor Mendel. In his day, people took much pride in developing what they believed were the best breeds of cattle, horses, poultry, and especially dogs and cats and other household pets. Community fairs and expositions were the scene of amazing displays of vegetables, fruits, and ornamentals that were the products of both amateur and professional horticulturists.

Many of these efforts relied on cross-breeding animals of the same species that differed from each other in important features, or cross-pollinating different varieties of plants. Mendel apparently had opportunities to observe this research and practice it in his monastery garden. "The striking regularity with which the same hybrid forms always reappeared whenever fertilization took place between the same species induced further experiments to be undertaken." Thus we see in the opening paragraph of his 1866 paper, "Experiments in Plant Hybridization," the curiosity that energized his investigations and ultimately launched the science of genetics.

While telling the O. J. Eigsti life story, I should point out that Mendel's paper was reproduced in *Principles of Genetics,* a textbook first published by McGraw-Hill in 1925. When Lysenko was deploring the sad state to which the Soviet genetics professors had brought their science, he cites this textbook as "an example of how uncritically our Mendelist-Morganists accept idealistic genetics is the fact that the standard textbook on genetics in many of our colleges has until quite recently been a translated American, strictly Morganistic, textbook by Sinnott and Dunn."

Mendel's experience as a gardener, together with a lot of good luck, led him to choose ordinary peas for his investigations:

> Some thoroughly distinct forms of this genus possess characters which are constant, and easily and certainly recognizable, and when their hybrids are mutually crossed they yield perfectly fertile progeny. Furthermore, a disturbance through foreign pollen cannot easily occur, since the fertilizing organs are closely packed inside the keel and the anther bursts within the bud, so that the stigma becomes covered with pollen even before the flower opens. . . . Artificial fertilization is certainly a somewhat elaborate process but nearly always succeeds. For this purpose the bud is opened before it is perfectly developed, the keel is removed, and each stamen carefully extracted by means of forceps, after which the stigma can at once be dusted over with the foreign pollen.

Calling on the seedsmen of his day, Mendel obtained thirty-four more or less distinct varieties of peas. He then grew each variety in exploratory plantings for two years. After sorting out the "few forms that were markedly different" from their own kind, he selected twenty-two varieties for his experiments. This collection enabled him to plant seeds that he knew would produce pea vines that grew no more than two or three feet tall. And he could choose between pea vines that produce white or purple flowers, pods that were flat or round, and peas that were wrinkled or smooth. There were additional equally clear-cut characters that other people had deemed sufficiently valuable to "select for" on the way to producing their favorite variety of garden peas.

Then came one of those critical quantum leaps in scientific thinking. Mendel must have realized that whatever was responsible for handing down clear-cut traits from generation to generation of garden peas had to exist exclusively in the pollen grains and ovules of the parent generation. Following this realization, we can see how he might have asked himself such questions as, "If this is the case, why is it so? How does it happen? What can I learn if I interfere with things as they happen ordinarily? What might be the result, for example, if the pollen grains that belong with a variety of very tall pea vines were to be exchanged with pollen grains from very short pea vines?" Similar curiosity could equally lead to pitting white flowers against purple, wrinkled seeds against smooth, and so on.

We take for granted that Mendel's years as a gardener taught him such details as how bees seem to prefer the early morning to transfer pollen from flower to flower, and that it would therefore be a good time of day to do hand pollination. Mendel probably was sure to snip off the anthers of the flowers that he expected to receive the competing pollen, and to protect their stigmas against receiving pollen from unanticipated sources. Also, he certainly kept accurate records. We can assume that Mendel was accustomed to letting the current year's pea vines mature and the pods dry so peas could be harvested for the seed for next year's crop. This was what he did with the peas from the vines that he had cross-pollinated. When these seeds were planted the following spring, the fun part of his investigation began. It was curiously exciting to find that the distinguishing character of one of the "parents" did not show up in the next year's vines and the peas they produced. It was as though that parent had been "out to lunch" when the pollination and fertilization processes were being carried out.

This puzzling state of affairs gave more reason to save the peas from these vines and plant them the following year, but, this time, to not interfere with the natural pollination process. These pea vines produced even greater excitement when characters that had been lost in the intervening generation reappeared. Since Mendel was also trained in mathematics, it is natural that he would keep track of the number of individual pea vines in this generation that had the distinguishing character of each of the original grandparents. For example, if one grandparent had been tall and the other short, he counted the number of short and tall vines in this third generation.

Another summer arrived and the plot thickened. For example, when the peas that grew on the short vines in that generation were planted, they produced only short offspring. Seeds from the tall vines, however, produced both tall and short offspring with the number of tall vines being several times greater than that of the short vines. (Mendel may not have realized how fortunate he was in selecting garden peas for his investigation. Had he chosen four-o'clock flowers, for example, and cross-pollinated red-flowered with white-flowered, the resulting offspring would have been pink, pretty much what would be expected and would therefore have not aroused curiosity).

The complete story of Mendel's work would take us through many years of cross-pollinating generation after generation of garden peas to observe how one trait is affected by being pitted against another. He gained highly informative data by keeping track of two or three traits during the same cross-pollinations—informative because he treated the data as being controlled by mathematical relationships. For example, using algebraic symbolism, one trait was designated by the letter "a" and another by the letter "b" and when together in pollen grain or egg cells, as "a + b." Cross-pollination is equivalent to multiplication and could be shown as $(a + b)(a + b)$ with the result being $a^2 + 2ab + b^2$. Translated back to the traits the symbols represent, this would call for one offspring to receive a double serving of the "a" trait, two to receive one each of "a" and "b" and one a double serving of the "b" trait. Keeping track of three traits at the same time would be shown as $(a + b + c)(a + b + c)(a + b + c)$ with the results being a rather awkward algebraic expression.

Missing from this telling of Mendel's story are elements that are both most important and most difficult to account for. We can only imagine the mental activity that translated the facts he derived from time spent tending and observing garden peas into a wonderfully satisfying explanation of how traits are handed down from generation to generation. But we do know that in 1865, he published such an explanation in the *Proceedings of the Natural Sciences Society of Brunn.* This is the paper that was left "on the shelf" for some forty years. Why this was so is pretty much a mystery. The usual explanation cites the obscurity of the journal in which the paper appeared.

There is a less politically correct explanation. Based on the most widely distributed translation, Mendel consistently refers to the thing that transmits traits from parents to their offspring as "characters." The term lacks the context of something real, actual, and particulate. As such, it generates uneasiness, as if something immaterial, mystical, or ethereal is involved. Nowhere in his paper does any term other than character appear. His messages become so much easier to grasp and become almost self-evident when we reach ahead four decades and substitute such a term as factor, or gene, or something that carries a material rather than ghostly context.

If we indulge in this substitution, Mendel's conclusions become beautifully evident. At the simplest level, each inheritable trait is controlled by two "factors," one from each parent. The factors for one trait act independently of those for other traits. Differences in the expression of a trait are due to differences in the controlling factor and when an individual receives two unlike factors for the same trait, one can be totally dominant over the other.

The height of pea plants illustrates this condensed version of Mendel's conclusions. Each pollen grain and each egg cell contains a tiny factor that can determine the height of the pea vine that would result from the fertilization of the egg cell by the male gamete in the pollen grain. If the two factors differ, one may be expressed fully while the other becomes "recessive" (but is not damaged in any way). Both factors will be passed along to offspring with neither being affected by the presence of the other in the cells of the parent generation. Each egg or pollen grain produced by the parent will stand an equal chance of containing either factor.

We come now to the type of mystery in the history of science that invites whatever explanation best reflects one's point of view. For more than three decades were there no reported efforts to describe more accurately the nature of Mendel's characters and how they functioned. As previously noted, the obscurity of the journal in which his work was published is the most common explanation. There is another and much more complex reason that reflects the scientific world's usual reaction to innovative points of view.

Something about the transmission of traits from generation to generation seems to invite resistance to new points of view, new ideas and theories, and new ways of doing things. There is more to it than protecting one's priority in scientific discoveries. Resistance can approach establishing taboos or shunning tentative hypotheses that threaten established beliefs or traditions. It is as though there are circumstances where ignorance is to be preferred over enlightenment.

Reactions to the work of Maude Slye in the twentieth century is a very poignant instance of the results from research in genetics not being followed up promptly. In her case, however, the investigations and results were announced in widely read publications. Maude Slye, like Gregor Mendel, spent much of her life studying the transmission of traits from parents to their offspring. Mendel cross-pollinated generation after generation of garden peas; Slye cross-bred generation after generation of mice. Mendel kept track of the inheritance of several traits; Slye concentrated on a single trait, namely, the onset of cancer in its various forms. It is most curious why both Mendel's and Slye's discoveries were left on the shelf for so many decades.

Slye kept as many as 10,000 mice in her laboratory at the University of Chicago. All received clean cages weekly. Each mouse was cared for daily, and when any mouse died, tissues and organs were inspected to establish cause of death. Early in the twentieth century, Slye began publishing the results of her investigations. Ultimately, she concluded that three factors, d1, d2, and d3, transmitted cancer (at least its beginnings) from parents to their offspring. The d1 factor caused epithelial cancer, the d2 factor caused cancers of connective tissues, and the origin of leukemias was associated with the d3 factor. The dominant counterparts of these factors which would prevent cancer, were identified as D1, D2, and D3.

Additional factors, together with their dominant counterparts, determined where cancers would develop. For example, the o factor located cancer in the ovary, the l factor in lung tissues, the m factor in the mammary glands, the b in the body wall, and the s factor in the skin. One mouse, for example, might inherit the pattern d1D1, d2d2, D3D3, L1, Mm, OO, Bb, ss. Since only the recessive d2 and s factors were not subjected to their dominant counterparts, this mouse could develop sarcoma of the skin. This mouse, in turn, would not develop, but could transmit to its offspring, susceptibility to carcinoma, sarcoma, or combinations of lung, mammary gland, and body wall cancer.

Neither Mendel's nor Slye's conclusions were followed up immediately by other scientists, but apparently for different reasons. In Slye's case, her conclusions evoked controversy even to the point of violently emotional rejection. Her methodology was attacked viciously and so was her integrity as a scientist. In rebuttal, she believed that subtle secondary interests, rarely openly admitted, were involved. Perhaps people were not yet ready to adjust to the emotional impact of learning that a predisposition to develop cancer can be passed from parents to their children.

Maude Slye believed that enlightening the public was the best hope for bringing cancer under control. In her mind, most people could gain the peace of mind that comes with knowing that the odds are heavily against their ever having any form of this horrendous disease. People who might inherit a predisposition to cancer deserved to know that early diagnosis of incipient cancers is often the difference between survival or death. Furthermore, it could well be that a lingering irritation or some other triggering event must occur before the genetic susceptibility became reality—an event that might be avoided if the victim was forewarned.

Thirty-five years after publication, other scientists happened upon Mendel's explanation of how traits are handed down from generation to generation. Some fifty years after Maude Slye began publishing conclusions based on her research, the public was told that it is "a certainty that we will soon come to fully understand" the origins of cancer. Much of this certainty was based on the recent discovery of "a cascade of genes, called oncogenes in cancer cells that appear to control growth and differentiation, whose misbehavior apparently leads to the aberrant growth we know as cancer."

Genetics illustrates beautifully how the pursuit of science cannot stand apart from all of the attributes that come with being human. Because science is creativity, inventiveness, and discovery, scientists share kinship with artists, composers, and poets. It is to be expected that they all strive to protect what they create, invent, and discover. But to do so poses problems in science that go beyond those of art, music, or literature. A painting, a symphony, a poem does not risk being tampered with by others who believe they can better execute what a prior work attempted to convey.

This is not so in the pursuit of scientific knowledge. Each glimpse of a bit of nature, each newly recognized interrelationship between nature's events or circumstances, each newly formulated scientific concept or principle invites being examined and reexamined by more penetrating ways and means of observing, and being tested for consistency with other discoveries in a similar domain. But the frailties of being human do not shield us from wanting to protect our stake in what we create or discover.

By no means is O. J. Eigsti an exception. On the other hand, O. J. represents all of the "little people" who put up with their share of the "slings and arrows" suffered by those whose life stories don't end up on the cutting room floor. Rejection of a new point of view, being told there is no future for one's new invention or way of doing something, failing to be recognized for one's personal discovery even of something long since known, are downers to be experienced often in everyday life. And the effects on the "little people" are every bit as devastating as for those whose life stories are widely publicized.

This has been very well documented in the case of Barbara McClintock, a contemporary of O. J.'s at the Cold Spring Harbor laboratory. Evelyn Fox Keller's *A Feeling for the Organism: The Life and Work of Barbara McClintock* (1983) enables us to feel vicariously the emotional impact resulting from having her creativity, inventiveness, and spirit of discovery be rejected repeatedly, or, even worse, tampered with by those who saw her work as putting at risk their own notions of how things should be.

Like O. J., McClintock was there while the science of genetics was struggling through its adolescence. Chromosomes were becoming recognized as having attributes far more significant than merely becoming observable after being readily stained with dyes. By the late 1930s, it was generally accepted that chromosomes played a key role in handing down traits from parents to offspring.

Many widely recognized geneticists accepted the "beads on a string" analogy to describe the relationship between genes and chromosomes. It was a convenient, comfortable analogy. It allowed each parent to hand down an equal number of the "family jewels." For those who needed such reassurance, each bead could be seen as having been enhanced by the achievements of its most recent owner or by being blessed by the advantages of a superior environment. This is only one of the many situations where the evolving science of genetics both influenced and was influenced by the public's beliefs and attitudes.

McClintock received her Ph.D. in botany from the College of Agriculture at Cornell University in 1927, nine years before O. J. earned his degree at the University of Illinois. She chose the genetics of corn as her area of specialization. Her biographer traces this choice to corn being of special concern at Cornell and "the colors of kernels on a cob of maize are a beautifully legible, almost diagrammatic expression of genetic traits." Throughout her life, McClintock seems to have managed to have her own field of corn as a very special laboratory.

Her uniquely effective skills and her genius as an investigator enabled her to see things happen in a corn plant and end up with a mental image of what had to have happened at the chromosome and gene level. This is where her troubled relationships with other geneticists began. Things that anyone could see, things as undeniable as speckles on grains of corn, simply could not have happened if genes were actually inviolate "beads" that could occupy only prescribed positions on a specific chromosome. Genes would have to change position on a chromosome. Equally disturbing was that there were "regulator" genes that controlled their position and "activator" genes that controlled their action. In the 1950s and '60s, these ideas were viewed as being totally unbelievable, and McClintock was treated accordingly.

It is essential that we consider a third and extremely significant episode in the history of genetics that unfolded concurrently with the life of O. J. Eigsti. Stripped down to the barest facts, this episode began when two scientists working independently came up with the observations that were ultimately "explained" as the actions of unbelievably minute organisms breaking loose genetic material from individuals of one species and allowing this genetic material to be carried to an individual of another species, where it would be treated as though it had always been a part of that specie's genome. Now known as genetic engineering, the managed transfer of genes from one to another species takes us to the frontier of the advances of genetics through the twentieth century.

The entire story of genetic engineering would more than fill this book. Dr. Donna Duckworth, in an article titled "History and Basic Properties of Bacterial Viruses," uses 20,000 words and 100 citations to convey the conception and early stages of the development of what she prefers to call "recombinant DNA technology." The article appeared in *Phase Technology* in 1987. Cited here are only the two sets of observations that coincided in time with O. J.'s early childhood.

In 1915, F. W. Twort was attempting to grow cowpox virus in the usual agar culture medium. The virus wouldn't grow, but he noticed that stray bacteria in the medium seemed to be afflicted with a disease. Watery areas around these bacterial colonies suggested something was interfering with the growth of the bacteria. Believing that perhaps an infective agent was involved, Twort collected the watery substances to see if he could filter out such an agent. The best explanation he could come up with was that the stray bacteria produced an enzyme that had the power to grow. This hypothesis, however, failed to stand up after further investigations.

At about the same time, Felix d'Herelle, while in Mexico, confronted areas where the ground was strewn with dead grasshoppers. He found they were infested with the locust coccobacilli, and he proved that these bacteria were killing the grasshoppers. This led to the idea of using coccobacilli to control invasions of grasshoppers. On one occasion, after preparing agar cultures of the bacteria, d'Herelle was surprised to find clear spots, very much like the watery areas Twort observed, where there should have been active colonies of bacteria. Microscope examination of the substances forming the clear areas revealed nothing. Furthermore, whatever was forming the spots would pass through a porcelain filter that held back bacteria.

These observations made by Twort and d'Herelle in the early 1900s, are of interest here for several reasons. They led to the identification of bacteriophages, the name given to what are now known as the viral agents that caused the clear areas in culture media. Explaining this action of the bacteriophage in turn yielded the first indication that the gene makeup of one species could be effectively changed by the invasion of a submicroscopic particle.

Today, for most people, genetic engineering evokes an image of overwhelmingly complex silicon chip–filled instruments and weirdly costumed scientists working in gleaming stainless steel laboratories on projects costing billions of dollars. They would be surprised to know that all of this started with the puzzling contents of Petri dishes and sick grasshoppers.

Equally surprising is the not yet widely recognized possibility that nature's "genetic engineering" may have vied with Charles Darwin's survival of the fittest in accounting for the origin of new species.

There is a point raised here. Why does it take so long for a scientific discovery to achieve fulfillment? Why so many decades between the first glimpse of a bit of nature and its ultimate satisfactory description? Why is there an even longer period before the new discovery is fulfilled by contributing to humanity's well-being? Why did Barbara McClintock have to wait some fifty years for her Nobel Prize? Why did it take even longer for seedless watermelons to advance beyond their initial invention to commercial success? And most poignant of all, why has it taken nearly the whole of the twentieth century for what Maude Slye learned about the origins of cancer to gain the recognition it deserves?

I raise these questions here for one reason. Whatever the answers are, they had much to do with what O. J. Eigsti and other twentieth-century scientists were able to achieve. Conjectural answers point toward the negative effects of being destructively attacked by one's colleagues as opposed to the encouragement that would come with constructively calling attention to possible shortcomings. Sincerely positive advice could have served well as an antidote to the toxic effects that came with making wrong decisions and dealing with unanticipated difficulties. Actually, O. J. recalls examples of both positive and negative interactions with those who were seeking the same goal: the commercial success of seedless watermelons.

Returning to an earlier metaphor, Eigsti and genetics pretty much grew up together through the period when chromosomes were the center of attention. Named initially because their most readily observed property was their special affinity for dyes and stains, these irregularly shaped objects that seemed to do mysterious things while cells were dividing, dominated the attention of geneticists well into the twentieth century. Spurred on by increasing assurance that chromosomes played key roles in handing down characters from parent to offspring, many scientists devoted their careers to describing in great detail what chromosomes were observed to do under both normal and abnormal conditions. Early on, discovery of the number of chromosomes that occur in a particular species merited publication. In time, even the slightest apparent discrepancy from the expected was sure to merit space in the literature.

When it became known that chromosomes carry still smaller objects, which we now know as genes, the latter took center stage. Much research was devoted to identifying where on each chromosome the gene for a particular trait was located. "Mapping" the location of all of the genes on each chromosome became a primary concern. The nature of a specific gene could be inferred by what it did. If, for example, a specific chromosome was proved to be uniquely related to the color of a person's eyes or to the color of a chicken's feathers, it was assumed that the responsible gene was on the specified chromosome. At the same time, because genes seemed to "cross over" from one chromosome to another, the frequency with which different genes on the same chromosome were "fellow travelers" between generations helped to establish their relative position.

During the "early adult" phase of genetics as a science, much attention was directed at relationships between and among genes that transmitted dissimilar expressions for the same trait. At the simplest level was the dominant versus recessive relationship first described by Mendel. This relationship among alleles, that is, unlike genes controlling the inheritance of the same trait, is often illustrated in school biology textbooks by the inheritance of eye color or, at a slightly more complex level, by the A, B, or O blood types. In reality, there is practically no limit to the complexity of the interactions of unlike genes for the same trait. To look at all of the characters of a species, and establish thereby the actual number of alleles and their relationships that control the transmission of each trait could approach being completely impossible.

One example is favored by textbook authors when they pose special challenges for their readers. The challenge begins by describing two totally white-feathered breeds of chickens. If these breeds are crossed, a predictable fraction of the "grandchicks" turn out to be brilliantly colored. Explaining this puzzling state of affairs calls for coming up with the relationship between alleles in which one kind of gene inhibits the action of a gene for a different trait. In short, one of the totally white breeds of chickens carries genes for color, but they cannot be expressed because this breed also carries an "inhibitor" gene. Furthermore, the inhibitor gene is dominant over the gene that permits the expression of color. Fortunately, students soon learn to use the traditional "checkerboard" system to see how normal gene segregation between generations permits "grandchicks" to inherit the recessive genes for both traits.

This is pretty much where genetics was when O. J. returned to the role of university professor in 1955. But already a huge leap forward was inundating this branch of science. In 1953, James Watson and Francis Crick added a new degree of depth in the struggle to describe accurately the transmission of traits from parent to offspring. Genes were no longer to be dealt with as giant molecules. Each gene was identified as a clump of atoms bonded together in a unique arrangement controlled by predictable chemical actions. In effect, each inheritable trait of any and all species could be traced to the action of a unique DNA (deoxyribonucleic acid) molecule (or the interaction of a few molecules that are sufficiently similar) to enable them to join in controlling the inheritance of the same trait but allowing the trait to be expressed differently.

O. J. now found himself as a geneticist living two lives. During the week, he was a professor presenting a textbook version of genetics to students in a totally urban setting. On weekends and during summers, he was a horticulturist practicing a chromosome and gene version of genetics in his watermelon test plots. By no means were the two lives without complications. Changes within the teaching profession were being driven by forces somewhat alien to academia. And in genetics, the shape of things to come suggested that O. J., Luther Burbank, and the generations of their colleagues who devoted their lives to improving the sources of food for the world's people had done things the hard way. In science, promises can play strange roles. At times, as with O. J., they can keep us struggling toward eventual success. At other times they can only lure us toward paths that never take us where we want to go.

This discussion of the role of hypotheses in the pursuit of science steps aside somewhat from the basic rationale of using the life of only one person as a window through which to view the study of genetics throughout the twentieth century. The question is why a hypothesis invites the attention of one person, but is rejected by another. The hypotheses underlying genetic engineering are good examples. They did not capture O. J.'s motivations. On the contrary, Lowell D. Owens sees the steps toward this technology as "historic milestones [that] foretell a future in which gene transfer will be used increasingly for the introduction of important agronomic traits into crops. . . . Increasingly in the future, gene transfer will be the method for introducing important, value-adding genes into elite lines of horticultural crops." (Dr. Owens is with the Plant Molecular Biology Laboratory, Agricultural Research Service, U.S. Department of Agriculture. This citation appears in "Overview of Gene Availability, Identification, and Regulation," in the August 1995 issue of *HortScience*.)

Conflicting interpretations of the role of genetics in humanity's well-being are expressed by Cheryl Long in the May/June 1997 issue of *Organic Gardening*. "Would You Like Some Weird Genes With That?" is the title of the article in which she argues against the "genetically engineered foods released upon an unsuspecting public by the 'gene jockey labs.'"

Dr. Owens and Dr. Eigsti received their advanced degrees from the University of Illinois. Both speak highly of their professors and the program of studies that led to their doctorates. Time may best explain the differences in the attitudes of these two scientists toward genetic engineering. When O. J. received his Ph.D., chromosomes were in the headlines. Some fifteen years later when Dr. Owens received his Ph.D., the newly revealed structure of DNA was of topmost interest, especially for those who were beginning their careers in science.

In addition to genetic engineering, there is today another rapidly developing field of research that could well affect the seedless watermelon industry. The fundamental concept is that each cell in the growing tissues of a plant carries a full set of that plant's genes. It is now possible to chop up these tissues into tiny fragments under conditions that keep the individual cells alive. In turn, each cell, or a small clump of cells, can be separated and nurtured to yield a fully developed member of that plant species.

When applied to the production of seedless watermelons, the tissue culture procedure begins with growing one triploid melon plant and chopping up the growing tips of its vines and roots. These thousands of bits of tissue are nurtured in greenhouses until they can be transplanted to become large fields of seedless watermelons. Bill Rhodes at Clemson University is glad to provide information about the tissue culture technology because he has two fond recollections of Dr. Eigsti. He notes,

> My first encounter, of course, was through the literature, and I was thoroughly versed on his role in seedless watermelon commercialization in this country when he called me in '81. A fire had gutted the building where all the seed of my breeding program had been stored. I lost everything but a few seeds kept in the lab. Dr. Eigsti said that he had heard of my loss and offered to provide me genetic material to start over. The other times I encountered Dr. Eigsti were at scientific meetings where we talked about—you guessed it—the seedless watermelon.

Rhodes went ahead to say that,

> Japan and Korea have developed and implemented the technology to graft large numbers of hybrid seedlings on disease resistant rootstocks of the same or different species. This technology will probably play a role in the mass propagation of seedless watermelon in the future. A number of American researchers have contributed to the cultural research necessary to advance commercialization of seedless watermelon: Adelberg, Compton, Gray, Zhang, and myself, and, of course, a number of researchers outside the United States. The tissue culture generation of diverse tetraploids, which can be used as parents of

> new triploids, is routine. Colchicine will probably be replaced by common herbicides as a chemical inducer of chromosome doubling in vitro. These herbicides are safer to the scientist, more effective and less detrimental to the plant material. If commercial cloning of triploid watermelon becomes commonplace, it will not be necessary to establish homozygosity in the parents, and individual hybrids can be increased to staggering proportions for the growers.

Rhodes shares with O. J. the same optimistic view of the future.

> Because the seedless watermelon still has great potential in the U.S. and the world, production will continue to increase. Our export market should grow because the seedless watermelon can be shipped around the world. The tenacity and ingenuity of Eigsti in 'growing' the seedless watermelon into a reality in the U.S. is understood and appreciated by all of the next generation of watermelon scientists.

I accept Rhodes's reference to "the tenacity and ingenuity of Eigsti." However, the very nature of the science of genetics, its ways and means as seen by the public, does not ensure that O. J. will be remembered, let alone "understood and appreciated." Seedless watermelons may become nothing more than a delightful product to carry home from the supermarket. The knowledge and know-how that made them possible will continue. O. J. Eigsti is proud to be a part of this.

The science and technology of genetics has become the equivalent of a boundless game in which the genomes of planet Earth's species are hands dealt by circumstances pretty much beyond knowable. To tamper with the genome of any species, even any variety of any species, risks being seen as invading a domain where we have no license to travel. In general, scientists are aware of this risk, but feel they cannot stand in the way of freedom to investigate all aspects of the genomes of each of nature's creations and how they got that way.

When O. J. Eigsti set out to develop a commercially successful variety of seedless watermelons, he joined the many scientists who were devoting their lives to understanding nature's ways of bringing together the specific genes that determine the characteristics, both anatomical and physiological, that set one species apart from all other species. His work on the seedless watermelon made him the dealer in a game where there could be as many as 10,000 different cards. He did not yet know the rules of the game, such as how one card was related to the others. Nor did he know for sure where he could obtain a full deck. For all he knew, the varieties of watermelons he had grown for years or could obtain from other seedsmen contained only some of the genes in the watermelon genome.

Much is being published about obscure relatives of other species being discovered throughout the world, particularly in tropical rain forests or other isolated habitats—relatives that carry uniquely valuable genes. It is little wonder that devotees of the genetic game want everything possible to be done to keep the genomes of all species complete and available. This game is difficult enough without having to take into account the possibility of key cards being missing from the deck. At the same time, it is the "tenacity and ingenuity" of the likes of O. J. that tell us few other games provide better opportunities to enjoy making one's life count for something even while flirting with failure.

In his book, *The Last Harvest: The Genetic Gamble that Threatens to Destroy American Agriculture* (1995), Paul Raeburn would have us believe that there are risks in tampering with the genetic makeup of any species, but particularly so for the approximately 100 species that provide most of the food for the world's people. In addition to the long known threats to potatoes, he cites epidemics that have swept through other species in which all varieties shared too many of the same genes. A spectacular example, the southern corn-leaf blight, first appeared in the United States in 1968; by 1970, destroyed 15 percent of the U.S. corn crop.

Susceptibility to this disease was traced to a single gene that had been built into hybrid corn seed being used in 85 percent of the corn acreage in the United States. Identified as the male sterile gene, its discovery was hailed as a major contribution to the hybrid corn industry because it replaced the costly, labor-intensive need to remove the tassels from one seed line to ensure cross-pollination. But the male sterile gene concept did not work out. In actual practice the presence of this gene interfered with the actions of other essential genes. And the hybrid corn seed industry had to return to sending busloads of workers into seed production fields, where they laboriously carried out the detasseling process. Although the production of seedless watermelon seeds only indirectly involves tampering with the genes that make watermelons what they are, valuable lessons have been learned from the male sterile gene episode in the hybrid corn seed industry.

It is difficult to use the commercialization of the seedless watermelon to predict "what's next" for this branch of science as the twentieth century draws to a close. We begin with questions. What did O. J. Eigsti have going for him that Luther Burbank did not have? What knowledge and know-how, what genetics-related gadgetry, equipment, and materials could O. J. take advantage of that were not available to Burbank? And at a less materialistic level, how did Burbank's friends and compatriots feel about him as a geneticist and what he was up to? Did these feelings resemble or differ from those that O. J. grew up with?

Put more directly, next spring when O. J. carries the seeds into his test plots in anticipation of harvesting an improved generation of watermelons, what does he know that Burbank did not know about what must happen in order to be able to cut a melon from the next generation and find its flesh to be sweeter, more crisp, a more attractive color than was the case for any melon in the previous generation?

The "what's next" spirit prompts even more challenging questions. While one watermelon seed is being replicated, what things continue to be every bit as mysterious to O. J. as they were to Burbank? What events and circumstances continue to be beyond our ability to describe, comprehend, or control? What questions do we not yet have answers for? What are the questions that have not yet been put into words?

Without doubt, the gene concept is at the center of things that O. J. had going for him that Burbank did not. For Burbank, Gregor Mendel's "characters" were nothing beyond a new, promising idea. None of the analogies we use to convey what the gene means to O. J. would have applied to Burbank. To associate the transmission of traits from parents to offspring with "informational molecules" could never have happened to Burbank. The fact remains, however, that both Burbank and O. J. were able to manage the transmission of traits in their experimental organisms.

Neither Burbank nor O. J. were to enjoy the advantage of knowing the true nature of the life entity. Life was known only by things that could be experienced. Until a potato was given a chance to sprout, until a watermelon seed was given a chance to germinate, there was no way to distinguish the living from the dead. No matter how detailed the description of something that could be observed along the way between the planting of a seed and the harvesting of a ripe melon, O. J. knew that everything depended on the watermelon plant being alive.

Early in the twentieth century, Burbank wrote, "My theory of the laws and underlying principles of plant creation is, in many respects, opposed to the theories of the materialists. I am a sincere believer in a higher power than man's." Late in the century, O. J. still believes that "there's something outside of us that keeps us going." As we come to the end of the century, there is an enormous accumulation of proclaimed "laws and underlying principles" that bear not only on the science of genetics, but on the nature of life itself. Many of these are at odds with each other. Some are overblown, given more credit than they deserve, or treated without due respect for interactions with other events or circumstances.

The mind-boggling conflicts that O. J. faces at the end of the twentieth century are well documented by Franklin M. Harold at the Colorado State University. He believes that genes as "information molecules" receive too much emphasis, and too little attention is given to what must happen before the "information" they convey is expressed in, say, the color and taste of a seedless watermelon. In language intended more for his peers, Harold says, "What never ceases to amaze me is how many scientists believe, or profess to believe, that genetic information is not only necessary but also sufficient to determine cellular morphology."

Here we see why the life of O. J. Eigsti may stand apart from, but at the same time, point toward "what's next" for bioscience in the twenty-first century. His place will always be in the watermelon patch, far from richly financed laboratories staffed with well-trained and dedicated men and women with access to state-of-the-art instruments. His role as guide-post is unique. He points toward genetics as it applies to life as a whole, to events and circumstances that are truly sufficient to enable a watermelon seed to be replicated—an entire seed, not a single physiological process or anatomical structure.

The capstone, the ultimate crowning event of genetics in the twentieth century may well be determining the chemical formulas, the two-dimensional word pictures of each of the "information molecules" that make up the human genome. Transferred to the story of the seedless watermelon, this is equivalent to having the blueprint for designing every kind of molecule that plays a role in a seed being replicated—each kind of molecule that appears during germination and the development of roots, stems, leaves, flowers, fruit, and the new seed. Each blueprint, at least indirectly, describes the kinds, numbers, and arrangement of the atoms that build one of the essential molecules.

In metaphor or analogy, publishing the complete watermelon genome would be equivalent to providing pictures of the parts needed to assemble a wonderfully intricate clock. But much is not provided. Nothing is said about procuring the necessary atoms to fashion each of the essential parts. Or the position of each of the thousands of atoms that make up many of life's essential molecules. Or how parts are joined together to make the clock tick. And most potentially challenging of all, our complete watermelon genome, the information needed to design each essential part, does not tell us how to start the clock running. Nor does it tell us how to include in the clock we assemble a full set of blueprints for tomorrow's clockmakers.

Chapter 5
In the Shadow of Luther Burbank

THE LIFE STORY of O. J. Eigsti and the seedless watermelon is interwoven with the evolution of genetics, horticulture, and education throughout the twentieth century. None of these enterprises started from ground zero at the opening of the century. Each was ongoing, pursuing its appropriate goals, continuing to develop ways and means for achieving these goals, and each was already deeply interwoven with the public's interests. In a very real way, these three enterprises offered a rich legacy for those who were to come after—a legacy that sustained O. J. in everything he set out to accomplish, though not always free from complications and contradictions.

This legacy is effectively delineated in the life story of Luther Burbank. Known primarily for developing improved varieties of many kinds of fruits and vegetables, his successes brought him local and international respect, even to the point of being accepted as a spokesman for the role science plays in making science and humanity mutually beneficial. His status as both a scientist and a citizen is best revealed in the twelve-volume collection of reports of his work that was published in a memorial financed by the Carnegie Foundation.

One of Burbank's biographers, W. S. Harwood, in *New Creations in Plant Life* (1905), lists many of Burbank's "leading creations," ranging from "the improved thornless and spiculess cactus, food for man and beast, to be the reclamation of the deserts of the world," to the "improved Australian star flower, one of the everlasting variety which is to be used for the decoration of ladies' hats."

Fruits were among Burbank's favorite targets. Among those cited by Harwood are "the primus-berry, a union of the raspberry and blackberry, the first recorded instance of the creation of a new species, together with the phenomenal berry created from the California dewberry and the Cuthberry raspberry; the plumcott, the union of the plum and the apricot, all three the accomplishment of what had been said to be an impossibility; a plum with no pit, one with the flavor of a Bartlett pear, one having a rare fragrance; blackberries without thorns; and the white blackberry, "a rare and beautiful fruit as toothsome as beautiful."

Harwood tells us that Burbank also "added much to the productiveness and edibility of vegetables." The examples cited include: "the pomato, an improvement on the poisonous potato ball, producing a fruit that grows upon the top of a potato" and "a variety of pieplant [rhubarb] with leaves four feet in diameter, bearing everyday in the year."

For flower gardens, there was the Shasta daisy, a blossom five to seven inches in diameter, made out of a wild field daisy; a Japanese and an English daisy; gladioli of greatly enhanced beauty, blooming around their entire stem like a hyacinth; a dahlia with its disagreeable odor driven out and in its place the odor of the magnolia blossom; a lily with fragrance of the Parma violet; and a scentless verbena given the intensified fragrance of the trailing arbutus; a poppy so increased in size that it measured ten inches across its brilliant bloom; an amaryllis bred up from two to three inches to nearly a foot in diameter; a calla increased in size until it measured ten to twelve inches in breadth, and then, the process reversed, bred down to less than two inches; and thousands of varieties of lilies.

In writing about O. J. Eigsti, it is important to mention that Burbank included detailed descriptions of his methods as his legacy to those who would continue his work. A favorite example was how Burbank developed "a plum which looks on the outside precisely like any other, but which . . . never fails to excite surprise. Even visitors who know what to expect when asked to bite through one of these specimens can seldom refrain from exclamations of wonder when the teeth go right through the fruit as readily as they would through a strawberry."

Before describing how he produced the stoneless plum, Burbank reminds his readers that "while it takes but a phrase to tell of the pollenizing of two plum flowers and the production of one anomaly in the first generation and of some other anomaly in the second, in reality a period of five or six years has elapsed between the pollenizing [sic] experiment and the observation of the second generation results. When this is borne in mind, and it is further recalled that breeding through many generations is necessary to secure the results desired, it will be clear that the production of a stoneless plum is an achievement that required its full share of waiting."

I quote Burbank at length for several reasons. First, His work is interesting in its own right. Even more important, it illustrates the spirit of exploration and discovery that was very much a part of our nation's environment during O. J.'s youth and early adulthood. He continues,

> At an early stage of my endless series of experiments in the hybridizing of plums, I chanced to hear of a so-called seedless plum that was said to grow in France, where it had been known for a long time as a curiosity. About 1890 I sent to the Transom Freres Nurserie in France and secured twigs of this plum, which was known merely as the Sans Noyau. These were grafted on one of my plum trees and in due course produced a crop of fruit, which as expected, proved to be a blue-black, cranberry-sized fruit, extremely sour, soft, and unfit for eating either raw or cooked. [As the plant grew] it became a rambling, thorny bush rather than a tree, utterly worthless for any purpose except the one for which I desired it. The fruit, besides being flavorless and palatable, was scanty in yield.
>
> Furthermore the fruit was by no means seedless notwithstanding its French name. It was only partially stoneless, as most specimens produced fair-sized kernels in the fruit, and every kernel had a thick rim of stone around one side partially covering the kernel. While it therefore lacked much of exhibiting the condition of stonelessness that I had hoped to see, it did, nevertheless, show a tendency to abandon the stony covering that has always characterized all of the fruits of the Prunus family.
>
> From the outset I was convinced that with proper hybridizing and selective breeding it could be made valuable. Next season the blossoms of the freak plums were fertilized with the pollen of the French prune and with that of numerous other plums and prunes. The seedlings from these crosses were grafted to ensure their earlier bearing. In the following generation I obtained some plums twice as large as their seed parent. Most of these had stones, however, and were soft, sour fruits. A very few of them were partially stoneless and from these the work was continued.
>
> The next generation showed some general improvement in the growth of the tree and in the size of the fruit. All the seedlings of the cross from the Sans Noyau upon the French prune were grafted and fruited, even though many of them showed the thorny, dwarfed, ill-shaped type of tree of the uncultivated ancestor. After two or three generations there was a

> marked tendency to improvement. In a large lot of seedlings in 1904, I obtained two that seemed to me of favorable appearance for much can be known from the quality of leaves and stem long before the time of fruiting.
>
> And when, two years later, the grafts thus selected bore fruit, it was delightful to find my predictions verified; the fruit was almost absolutely stoneless only the faintest splinter of stone occasionally appearing. And combined with this seedless condition, there were qualities of size and flavor that made the fruit practically equal to the French prune. Moreover, as is often the case with hybrids one strain of which is wild stock, the new plum proved to be a very good bearer. So my ideal of an eatable plum having no stone about its seed was almost achieved.
>
> I say almost achieved because there still remained, in the case of the plums of best quality, a fragment of shell which varied from a small crescent about one side of the kernel to an almost invisible granule. There were some individual plants among the numberless seedlings that bore fruit in which the stone was absolutely eliminated and, in some cases, the seed also. But it proved extremely difficult to combine the quality of entire stonelessness with the desirable qualities of size and flavor, lacking which the fruit could have no practical value.
>
> Further hybridizing experiments, aimed at the production of an absolutely stoneless plum of fine flavor, are still under way; but in the meantime there are several varieties actually in hand that are of most admirable quality and yet stoneless. . . . And among the eight or ten hundred varieties of stoneless plums now growing in my orchard there are sure to be some that will show still further improvement.

It is always difficult to trace any scientist's discoveries and the strategies that made them possible to his or her notions of nature's fundamental principles and ways of doing things. This is especially problematic when new points of view are struggling to be recognized. It is significant that Burbank had already enjoyed success in his work before Mendel's "unit characters" became common knowledge. It might have been that Burbank didn't need knowledge of Mendel's discoveries and could easily have resisted incorporating them into his beliefs and strategies. On the contrary, we see him adopting Mendelian terms when he says, "We discover soon that the stone seed is prepotent or dominant and stonelessness latent or recessive. So we must be prepared to see the progeny of our first generation of hybrids all produce stony fruit. But a knowledge of latent or recessive characters to reappear in successive generations comes to our aid, and we go on with the experiment with full confidence, even though for the moment we seem to be going backward rather than forward."

Eigsti pretty much escaped the intellectual stress of having to give up one notion and accept a new comprehension of how traits are inherited. By no means, however, has he been able to avoid almost continuously adjusting his beliefs and investigational strategies to accommodate new discoveries and insights. There is another, and very important lesson O. J. may well have learned from the Burbank legacy. Burbank's biographers consistently see him as clinging stubbornly to the belief that acquired traits can be handed down from generation to generation—stubbornly, even though he says, "All scientists have found that preconceived notions, dogmas, and all personal prejudice must be set aside, listening patiently, quietly and reverently to the lessons one by one which mother nature has to teach, shedding light on that which before was a mystery, so that all who will may see and know."

What we are talking about here is expressed by David Starr Jordan in *The Scientific Aspects of Luther Burbank's Work* (1908):

> Within the range of his own experience he is an original and logical thinker, and his conclusions are in general most sound. He is not a physiologist, still less a histologist, and the phenomena of heredity as shown in cell-multiplication, he has not studied for himself. The researches of Weismann and those suggested by his theories of heredity Burbank has given little attention to, and he has, therefore, a confidence in the inheritance of acquired characters, such as effects of environment, which most biologists of today do not share.

Burbank's beliefs show through such statements as the following:

> Varieties are the product of fixed laws, never of chance, and with a knowledge of these laws we can improve the products of nature, by employing nature's forces, in ameliorating old or producing new species and varieties better adapted to our necessities and tastes. Better food, more sunshine, less arduous competition, will of themselves induce variation in individual plants which will be more or less transmitted by their seedlings, which, selected consecutively through a certain number of generations, will become permanent. Environment here exerts an influence as in all chemical, cosmic, and celestial movements. These small increments from environmental forces may produce a gradual or sudden change according to circumstances.

Another of Burbank's beliefs can be seen as adding to his stature as a scientist. It would be reaching to suggest that he anticipated today's emphasis on DNA as the continuing linkage within all forms of life, but the following passages give cause for thought.

> Every tree in the orchard, for example, carries within its tissues a portion of protoplasmic matter that has come down to it through an almost infinite series of growths and divisions in unbroken succession from the first tree that ever developed on the earth—or, for that matter, from a vast series of more primitive organisms that were the progenitors of the first tree.
>
> And while this stream of primordial protoplasm has been changed by an infinitesimal quantity in each successive era, it has retained even to the present the fundamental characteristics that it had from the outset. That such is the case seems little less than a miracle; that an almost microscopal speck of protoplasm which we term a pollen grain should contain the potentialities of thousands of generations of ancestors, and should be able to transmit them with such force that the seed growing from the ovule fertilized by that pollen grain will inevitably produce, let us say, an apple tree, not a pear tree or a plum, is beyond comprehension.
>
> Yet we know it to be true. And so the plant hybridizer who consciously merges two different protoplasmic streams when he brings the pollen from one flower to the pistil of another, participates in what must be considered the most wonderful of all experiments.

There is one more element in the influence that Luther Burbank had on the lives of scientists who grew up with the twentieth century. His widespread acceptance by so many facets of his society apparently gave him license to interlace philosophical beliefs with his interpretations of the ways and means of nature. We can only wonder how people of his day, particularly young people who were beginning their careers in science, would have been affected if they contemplated:

> My theory of the laws and underlying principles of plant creation is, in many respects, opposed to the theories of the materialists. I am a sincere believer in a higher power than man's. All my investigations have led me away from the idea of a dead material universe tossed about by various forces, to that of a universe which is absolutely all force, life, soul, thought, or whatever name we may choose to call it. Every atom, molecule, plant, animal or planet, is only an aggregation of organized unit forces, held in place by stronger forces, thus holding them for a time latent, though teeming with inconceivable power. All life on our planet is, so to speak, just on the outer fringe of this infinite ocean of force. This universe is not half dead, but all alive.

These discussions that go beyond Burbank's work in gardens, fields, and orchards point toward some of the more complex features of what it has meant to O. J. Eigsti to grow up with science during the twentieth century. One such feature is related to the Lysenko affair, when we read, "Better food, more sunshine, less arduous competition, will of themselves induce variation in individual plants which will be more or less transmitted by their seedlings, which, selected consecutively through a certain number of generations, will become permanent," Burbank apparently unknowingly anticipated by some four decades the fundamental platform in the political agenda identified as Lysenkoism.

But his name does not appear in the previously cited *Proceedings of the Lenin Academy of Agricultural Sciences of the U.S.S.R.* Many examples of Burbank's kind of research when done by Soviet agroscientists are discussed in detail. Improvements in dozens of farm products receive unlimited praise. The success obtained by this kind of research "proved," as proclaimed by Lysenko and his followers, that acquired characteristics can be inherited. Many paragraphs also "proved" that the leading geneticists in America were wrong in limiting the transmission of characteristics to the action of genes. But, again, no mention of Luther Burbank.

Based on the total of Burbank's beliefs and knowledge, the reason can be no more than conjecture. For sure, there would be no place in a nation dominated by Lysenkoism for "a sincere believer in a higher power than man's." Or, as we recall that sermon O. J. preached so many years ago, for someone who believes "there's something outside of us that keeps us going."

Finally, the example of Luther Burbank gives us cause to wonder about how the future will deal with O. J. Eigsti, what he has done, and what he stands for. It is curious to see that one of Burbank's creations, the Burbank potato, was only briefly mentioned by his biographers. Dozens of other successes received far more attention. Yet it is the Burbank potato that has kept his name alive throughout the twentieth century. Each time a McDonald's customer nibbles a French fry, it may well have come from a direct descendant of potatoes that once grew in Luther Burbank's gardens. And thereto hangs a tale that must be told.

The Russet Burbank is the principal variety that McDonald's contracts with potato growers to produce. What we once thought of as chiefly an Idaho product has now taken the Burbank name around the world. The characteristics that Burbank bred into this potato make it the variety of choice for the French fry industry. It is ironic that this lack of genetic diversity can also be seen as creating a threat to all potato growers and their financial well-being with a lack of genetic diversity.

Reliance on a single variety of an essential crop was the cause of the devastating famine that Ireland suffered halfway through the nineteenth century. Because potatoes flourished in Ireland and provided seemingly unlimited and nutritious food for everyone, they soon became the dietary staple for a very large fraction of Ireland's 8 million people. One variety, the lumper, was the best of all varieties. In time, not only was Ireland's food supply dependent on a single species, it was heavily dependent on a single clone, that is, a variety whose members were all genetically identical.

A second factor contributed to the devastating Irish famine. Traditionally, varieties of food-producing plants and animals that differ in their genetic makeup rarely are equally susceptible to the same disease or insect infestation. An endemic disease or widespread insect invasion may wipe out one, or, at most, a few varieties, but not the total species. Early in the 1990s, a new variety of Phytophthora infestans, the same blight fungus that caused the Irish famine, began showing up in potato fields in British Columbia and Pennsylvania. The Burbank potato's genes were unable to cope with this variety of fungus. Growers began suffering enormous financial losses from fields that sickened and died before marketable potatoes could be harvested. To add to the problem, this newly mutated variety of the fungus was resistant to all currently acceptable herbicides. Potatoes grown in fields where the fungus was only partially established did not escape damage. Huge quantities of potatoes in warehouses rotted before they could be sold. Losses to large-scale growers reached well into six figures.

Of course, I don't suggest that this state of affairs should be charged against Luther Burbank's contribution to humanity's well-being. But this kind of result happens often enough to be of concern to all scientists, particularly to those whose investigations involve genetics. And O. J. Eigsti is no exception.

Chapter 6

O. J. as Author: The Colchicine Book

WRITING A BOOK isn't something one might expect from a man whose main interest in life was the development of a high-quality, commercially successful seedless watermelon. Long hours spent tracking down elusive references in sometimes distant libraries is quite different from studying thin slices of watermelon plant tissues under a microscope to determine the number and condition of the chromosomes or making weekly visits to farmers' fields to observe which watermelon plants are best able to fend off insects or avoid falling victim to disease.

The question is how did O. J. Eigsti develop the inescapable urge to share with others his interest in all of the things colchicine can do in agriculture, medicine, biology, and chemistry? One quite logical answer (or at least a partial answer) takes us to a feature of higher education that may very well have been conceived primarily to encourage scholars in all intellectual endeavors to make the knowledge they create available to their colleagues. To write a scholarly thesis has long been required for candidates for advanced degrees at the world's universities—a requirement that is as likely to generate as many negative as positive responses on the part of the candidate. The contents of master's and doctoral theses are not always held in high regard by those who pass judgment on such matters.

"A Contribution to the Life History of Resede odorata L,." and "A Morphological and Cytological Investigation of Certain Members of the Resedacese," the titles of O. J.'s master's and doctoral theses could well be cited as examples by those who see such investigations as being "meaningless efforts to learn more and more about less and less." At the same time, we can read in his colchicine book such statements as "since the senior author had been privileged to attend seminars in cytophysiology by Professor C. F. Hottes, University of Illinois, the polyploid cells found in treated root tips at the Carnegie laboratories received more than average passing attention." Here is O. J.'s comment: "That's very important. That must absolutely be stated. If I had not been exposed to Hottes' work on chromosomes and heat treatment and arrest of mitosis at the metaphase stage, I don't think I could have hit upon the colchicine effect as I did."

Although his friends were well aware of O. J.'s compulsion to do the colchicine book, its actual creation did not become possible until a set of remarkably random coincidences occurred. O. J. tells the story:

> *I got the idea of doing a review of the uses of colchicine. I wasn't thinking of doing a book. I went up to Douglas Lake in Michigan in 1943. The idea was to write this review and take it back to the University of Oklahoma because they wanted to publish a book on colchicine. Well, I went to Douglas Lake, spent six weeks, then went back to Oklahoma. The war was on and I got involved in corn work. Then at the end of the war, I published in monograph style this bibliography for all of the papers that had been published on colchicine.*
>
> *Somehow Dustin [Dr. A. Pierre Dustin Jr.] got hold of a copy. It was distributed from the Oklahoma University Press. We didn't meet until he came from Belgium in 1949 to attend the International Congress on Cancer at St. Louis. And he came up to Evanston to visit us on his way back to Brussels. Then he told me about his father working on this review of colchicine. His father died during the war, of a heart attack while he was held as a prisoner. You see, the Nazis wanted him to open the university [University of Brussels], he was president of the university, under Nazi direction and he wouldn't do it. So they plunked him in prison. He was in prison when he wrote the manuscript. He wrote it in French. When we were working together in Brussels in 1950, Dustin gave me this manuscript. I still have it.*
>
> *We got together. He would do the chapters on medical and I would do the chapters on cytology. I would do the work on agriculture, botany, polyploidy, and so on. We got it all together. I went to the International Genetics Conference in Sweden and I met a geneticist from Iowa State. Somehow I found out that he was chairman of the committee that accepted manuscripts for the Iowa State University Press. Then I went to Brussels for a couple of weeks and we finished the manuscript. I then sent it to Iowa State Press and they accepted it and were going to publish it.*
>
> *Then I got the Fulbright. I wrote to the editor and told him I would be going around the world and would be visiting many laboratories where colchicine work is going on. It would be terrible to*

> *publish this book without including all of the latest information. Please send the manuscript back to me. And he did. So I carried it with me all around the world. One night I was reading this manuscript and I said this just doesn't sound right. The language wasn't right, the phrasing was awkward. It just wasn't right. What should I do? I took the manuscript, we had a fireplace going, and burned it. Didn't save a word! I had already submitted it to the Iowa State Press. But I burned the thing*
>
> *The next day I began rewriting the manuscript by hand. I wrote a paragraph a day. Agnes typed every day. I did it word by word. Blood, sweat, and tears. I talked to Dobzhansky at a meeting and I said, "Dr. Dobzhansky, I read your book when I was in Pakistan and I really admired the way you made things stand out, everything was easy to read, and so clear. You must just sit down and write." "Oh," he said, "you're wrong. It takes me sometimes a whole day to get just one paragraph written." I said I'm sure glad to hear that because that is what happened to me.*
>
> *So I got to Brussels and Pierre began reading the manuscript. He said, "this isn't the manuscript we sent to Iowa State." No, I burned that. We almost split. It took two weeks before we were back together. Agnes kinda kept peace so we didn't kill each other. Finally one day he said, "I've read the manuscript, I'll stay with you." So I spent the rest of the summer in Brussels. The book was eventually published. When Pierre got a copy he wrote me the nicest letter. He said, "I am so pleased because if my father could see this work he would be very pleased." We dedicated it to his father, you know.*

Out of respect for all authors who have shared in these experiences, the contents of this letter, which was sent from Brussels to O. J. on January 8, 1958 are quoted here:

> Dear O. J.: A few minutes ago the moment I was expecting since so many years happened: there was the book on my desk. Everything seems perfect, binding, typography, paper, figures, and so on. This copy will be cherished by me for its dedication, and once more I feel very grateful toward you, without whom "Colchicine" would have never been anything more than a dream.
>
> I feel deeply convinced that my father—somewhere in the Universe—must feel happy that the task he began in the dark years of the war has come to completion. Many thanks are due to Agnes—not only for all her work, but also for that fine hospitality she gave in 1949 to this difficult Belgian doctor!
>
> So dreams may come to life if only one persists in the same direction. Hoping you are rejoicing, as I am, dear O. J.

Your sincere friend, Pierre.

This bringing together of deeply felt motivations had much to do with overcoming the obstacles that stood between the desire to publish the colchicine book and its ultimate realization. But there are also important things to deal with if this book is to portray O. J. Eigsti's life as a slice of the whole of the twentieth century. This book concerns one of the most basic of life's mysteries, something that surely spurred the interests of its authors in ways that saw them through the demanding chores that are essential to the author's craft. But these interests stem from exposures to nature's phenomena that began near the beginning rather than the middle of the twentieth century.

Colchicine influences the life process at the crucial point where one cell becomes two cells. With all forms of life beginning with a single cell, no process other than life itself can be more significant—or more of a challenge. O. J.'s first formal ventures into the pursuit of science involved the cell division process—ventures prompted primarily to fulfill the requirements for advanced degrees. Perhaps it was the fundamental importance of this topic that set him apart from those who remember master's and doctoral theses as nothing more than obstacles to be overcome.

O. J. completed the research for his graduate degrees during the early 1930s. At that time, scientists were particularly puzzled by what they saw when their microscopes were focused on plant and animal tissues that had been smashed flat or sliced very thinly and stained with special dyes. A favorite tissue was the tips of onion roots. An onion left in water for a few days is almost sure to produce roots, and it was common knowledge that roots grow primarily from their tip ends. At that time there was no way to observe living cells actually dividing. The best that could be done was to study slides in which the division process was "frozen" in time.

A standard lab exercise for biology students called for sketching and labeling the arrangement of oddly shaped dark objects their microscopes enabled them to observe in the otherwise almost transparent onion root tip cells. Lab manuals advised arranging their sketches in a special sequence. What they saw on the glass slide on a microscope stage was not always associated with mitosis or any other part of a cell's life stage. Only rarely were students enabled to observe a live onion actually doing what the day's lesson was expected to teach. For all too many people, this lab exercise never became anything more exciting than dutifully sketching and labeling "oddly shaped dark objects."

But not for O. J. Eigsti. In time, those "oddly shaped dark objects" became the equivalent of thousands of ballet dancers appearing from nowhere and responding to a choreography communicated from wherever and whatever it means to be alive. Each cell in that onion root tissue revealed wonderfully disciplined dancers frozen in time and space.

Obviously, there was no mention of ballet dancers or choreography in the research journals of the 1920s and '30s. Mitosis and meiosis involved chromosomes responding to a totally effective albeit equally mysterious growth-causing chemistry. In time, it was learned that the number and shapes of these oddly shaped dark objects were the same for all members of the same species. But in some cases, individuals would have two or more times the number of chromosomes in their cells. Where there were differences in the shape of one or more chromosomes, there were likely to be discrepancies in the age-old concept that "like begets like." O. J. wants it known:

> *If we're talking about the early history of chromosomes, one man must be mentioned. It was John Belling who developed the acetocarmine technique whereby whole chromosomes became visible. Belling worked at Cold Spring Harbor and without his discovery of this technique, much of the work of his more widely recognized colleagues would never have been possible. The acetocarmine technique changed the whole picture. No longer did we have to rely on slicing chromosomes very thin. All we had to do was to smash plant tissues, add the dye, and the chromosomes practically jumped out at us.*

In time, it became increasingly evident that chromosomes played key roles in the transmission of traits from parents to their offspring. How this realization was shaped and became accepted is a good example of the resistance people must deal with when they present new points of view regarding the handing down of traits from generation to generation. What is common knowledge today is the final residue left behind by people who pursued sometimes highly controversial hypotheses to their ultimate proof or disproof. But it isn't always possible to give credit where credit is due.

It is to be expected that O. J. would be caught up in this interest in chromosomes when considering possible topics for a graduate level thesis. When the chosen topic became polyploidy, the phenomenon in which individuals show up with two or more times the usual number of chromosomes, no one could have foreseen this might be the initial step along the path that would take him not only to writing a book, but to becoming committed to the life-long mission of providing high-quality seedless watermelons.

Actually, the urge to publish a book and share his colchicine-related research with others gained momentum while he was working as a research associate at the Carnegie Institution of Washington's laboratory at Cold Spring Harbor on Long Island. It all began with an idea shaping up while O. J. was attending a staff seminar devoted to the effects of colchicine on mitosis, specifically, the migration of the newly replicated sets of chromosomes to their respective "daughter" cells. Not always was it only a single set of chromosomes that migrated to each of the newly forming cells. O. J. saw this as a possible way to cause polyploidy, that is, to produce new varieties of a plant by causing its cells to contain two or more times the usual number of chromosomes.

Excited by the possibilities that existed in this idea, he obtained a supply of colchicine before he left the seminar. Quoting from the story as he tells it in the colchicine book, he:

> Obtained colchicine for Alium root tip tests before leaving the laboratory. Appropriate concentrations were determined for the experiment with plant material. Within 72 hours, large bulbous tips appeared on onion roots immersed in colchicine; the cells showed polyploid restitution nuclei by acetocarmine methods. Since the senior author had been privileged to attend seminars in cytophysiology by Professor C. F. Hottes, University of Illinois, the polyploid cells found in treated root tips at the Carnegie laboratories received more than average passing attention.

When O. J. Eigsti was a neophyte among his well-established colleagues at the prestigious Cold Spring Harbor laboratory, he felt he deserved full credit for discovering that colchicine can be used to cause polyploidy in ways that can be particularly advantageous. When he was denied this recognition, he abruptly left the Cold Spring Harbor laboratory—a laboratory long known as the spawning ground for scientists of Nobel Prize rank. Furthermore, the nation was beset by the Great Depression at the time. Jobs for scientists, both the newly prepared and the experienced, were very scarce. O. J. had only the promise of a part-time job with Dr. George H. Conant, whose principal business, Triarch, was preparing and distributing instructional materials for the biological sciences.

Memory of the Cold Spring Harbor incident became all the more agitating when what O. J. deemed to be his personal "discovery" was published in a French journal by a senior member of the Cold Spring Harbor staff. Fortunately, in keeping with the long-standing tradition of faculty advisors retaining interest in the professional careers of those whom they advise, Eigsti's advisor at the University of Illinois made it possible for him to publish "A Cytological Study of Colchicine Effects in the Induction of Polyploidy in Plants" in the 1938 issue of the *Proceedings of the National Academy of Science*. It is true that this action did not reestablish priority for what was being published, but it was certainly an appreciated leg up for O. J. in the area of learning to take in stride and ultimately benefit from the many attributes of the science profession, both the positive and the negative. This incident at Cold Spring Harbor may well have become the burr under O. J.'s saddle that saw him through the commitment to publish the colchicine book.

O. J. Eigsti was still in his twenties when he confronted this traumatic career decision, more traumatic in memory than in reality. At the time, he believed he was being mistreated by the scientific "establishment" in ways that gave him no recourse. Looking back through the six decades that followed, however, we see a remarkable instance of an individual who managed to both benefit from and contribute to the science community, with all of its complex ways and means, while sometimes being within and other times being apart from the "establishment."

O. J. has been free to pursue research of his own choosing. Much of what he has done is highly specialized. At the same time, his work on the seedless watermelon has yielded a multimillion-dollar capstone to a wonderfully multifaceted career. We can only regret that his Cold Spring Harbor experience was tinged with feelings of resentment forever after.

To put this resentment into perspective, begin with O. J.'s introduction to the heady atmosphere of the nation's most prestigious genetics research laboratory. It is the summer of 1935. He is to work with his doctoral advisor at the University of Illinois, who is among the many high-level scientists who always looked forward to spending the summer at Cold Spring Harbor. The adjustments awaiting O. J. can be best appreciated by beginning with his way of traveling from rural Illinois to Long Island.

> *I had a special train ticket from Champagne to Niagara Falls. When I got to the Falls, I gave them a quick look and decided to continue on to New York. My traveling companion took a bus to New York City but I decided to hitchhike. I had no idea how far it was and got on the road about five o'clock in the evening. I wasn't on the road long until I was picked up by a man in a Ford roadster, about 1925 vintage, no windshield, no top, and a pretty rough looking driver. He was surprised that I didn't know how far it was to where I wanted to go but he said he would take me to Albany. That took us all night.*
>
> *The moon was full so it was a wonderful night. Warm enough to be comfortable and he didn't drive fast enough to make the lack of a windshield a problem. He turned out to be a fine companion. When we stopped for coffee, he insisted on paying. He assumed I was not very well heeled. And when we approached Albany, he stopped to shave before meeting his family. That was a ride I will never forget. And I got to the Greyhound bus station in New York before my friend did! But I took the train the rest of the way to Cold Spring Harbor.*

We move ahead to February 1937 at the Department of Genetics of the Carnegie Institution of Washington, where O. J. was paid $110 per month as a full-time employee.

> *My laboratory is in the building called the animal house where mice, rats, and pigeons are kept for genetics studies. This is the building where some of the world's best known cytologists were working on the giant chromosomes in the fruit fly. One of my duties was to x-ray male flies. I particularly remember Dr. C. B. Bridges. He was the greatest scientist I have ever known. He died at age 49.*

More than geography separated O. J. from familiar associates. He was among graduates from universities whose traditions were not those of a midwestern land grant university. There were other problems. The Eigstis lived in a tent, scarcely appropriate for newlyweds. At the same time, their associates came from and would go on to rise to high places in both science and society. The stories O. J. and Agnes still tell suggest a rare ability to continue to be the people they were in their Mennonite community, but not to feel alien among those with widely diverse backgrounds. A delayed honeymoon trip to New England includes a typical "adjustment."

> *We borrowed a camping outfit, tent, cots, etc., and set out for Vermont. At one place we tried to buy a chicken. The only possibility was to buy a live one, dress it, and cook it over the camp stove. It was delicious. After Vermont we touched New Hampshire, then went down the coast to Boston. We asked the police about a place to set up our tent for the night. He suggested the Boston Common. I only wish we had a picture to document that camp!*

The diversity of the Eigstis' circle of friends while at Cold Spring Harbor is noteworthy. At least one of their "tent neighbors" became a Nobel Prize laureate. Then there was the butler to the owner of one of the large estates nearby. When the family was away, he would entertain the Eigstis and share with them the joys of wealth. Striking a different note, the butler would bring left-over flowers that carried the bittersweet impact of being selected from party decorations that cost more than O. J.'s annual salary. One of Agnes's special friends stopped over while returning from visiting family in Germany and told of being mistreated for not raising her arm in the Heil Hitler salute. Her stories about what was happening in Germany were ominously foreboding.

All of this is to say that O. J.'s decision to leave Cold Spring Harbor was not spawned from a single event. It may well have been more environmental; no matter how traumatic it seemed to be at the time, the disappointment stemming from feeling that his work was not receiving due recognition should be weighed against the opportunities to rub elbows and minds with singularly successful people.

There is only one reason to dwell on the Cold Spring Harbor incident: the pursuit of science cannot escape being a group activity. At the same time, creativity and inventiveness are highly individualistic attributes of the human spirit. Group dynamics, the ways and means whereby the group affects the individual, are not always in the best interests of the individual. Those who are compelled to live as individuals can easily find themselves at loggerheads with those who are adept at practicing the tactics and strategies of group dynamics.

It is interesting that the experience of one of O. J.'s colleagues during the Cold Spring Harbor incident has been repeatedly cited as an example of what is being discussed here. Barbara McClintock's biographers have had much to say about her being put down as a person by far too many members of the scientific community. From the 1930s through the '60s, the results from her investigations were virtually ignored. During the 1970s, things took a different turn. Her discoveries simply became impossible to reject. When she received the Nobel Prize in 1983, Barbara McClintock became the role model for young scientists of both genders whose devotion to their profession can be safely tested by what are too often the ways and means of those who put group dynamics above the interests and emotions of the individual members of the group they are in position to control.

Colchicine – In Agriculture, Medicine, Biology, and Chemistry, published by the Iowa State College Press in 1955, is a 470-page study that cites 1,652 investigations. It is also much more than a way for two individuals to fulfill the obligations that come with being scientists. At a far more poignant level, it is a son's memorial to a father who was among the millions for whom the twentieth century brought utter misery rather than fulfillment.

The contents of this book continue to bear on problems that are not yet solved. The copy in the library of the National Institutes of Health shows evidence of much use, and so does the copy at the National Agricultural Library. This suggests a philosophical comment on research people, particularly those who focus on improving the quantity and quality of food and fiber that becomes available worldwide. Their names will rarely, if ever, be associated with the magnificent display of fruits and vegetables in our supermarkets or the world's supply of other agricultural products. But the time and energy they spend sharing their knowledge and know-how promises to yield very special rewards: Books are archival, doubly so when their contents deal with the urge to improve the quality of life for the world's people.

The index of the colchicine book is a veritable catalog of the physicochemical aspects of what it means to be alive. There is much tracing of cells dividing to form tissues, tissues becoming integrated to form organs, and organs performing the myriad functions of life's countless forms. When the book was written, the focus was on chromosomes. Today, the focus is on genes, specifically the unique forms of the DNA substance that code for each of the kinds of molecules that make up the chromosomes they now replace on center stage.

Something else gives this book special value, particularly for science teachers and their students. It presents the tactics and strategies, the ways of science as they were practiced when science was advancing beyond its adolescence. Episodes in science are seen as beginning with observing an event in nature that is out of the ordinary, something happening that arouses curiosity because it contradicts what is expected. In turn, this curiosity leads to observing the nonconforming event from every possible angle in the hope of catching a glimpse of why it happens. The most promising explanation becomes the first working hypothesis.

Hypotheses, in turn, call for verification, usually by carrying out carefully designed experiments. At their simplest level, experiments employ such strategies as (1) proving that every time the event in question happens, the assumed cause is present; (2) proving that the event cannot occur when the assumed cause is absent; or, (3) proving that a change in the magnitude of the assumed cause is always accompanied by either an increase or decrease in the magnitude of the event that is being investigated.

Polyploidy, the state of affairs in which the cells in an animal or plant's tissues have two or more times the usual number of chromosomes, is obviously out of the ordinary. How O. J. and his coauthor describe the role of colchicine in causing this "disconformity" becomes a great example of science in action. First, they devote thirty-one pages to a review of other people's efforts to describe the mitosis process as accurately as possible. Special attention is paid to the stage in cell division, metaphase, where the replicated chromosomes separate and move to the new daughter cells.

This special attention begins to focus on the spindle fibers, the structures that appear during mitosis and seem to pull or otherwise cause the replicated sets of chromosomes to migrate to their respective daughter cells. Here is where an hypothesis begins to take shape. Quoting Eigsti and Dustin: "If the spindle fiber is the substrate where colchicine acts—and there are many data to support this assumption " Then cytological and biochemical methods should show us more clearly what reactions occur. . . . Provisionally, let us say that colchicine alters rather than totally destroys the spindle substance."

It is relatively easy to design experiments to prove whether or not the presence of colchicine during mitosis results in changes in the spindle fibers that are adequate to cause polyploidy. It is equally easy to prove whether or not changing the strength of the colchicine solution is accompanied by changes in the degree to which the spindle fibers are affected and, hence, in the number of cells that acquire more than the usual number of chromosomes. Because polyploidy occurs in nature, however, it is not possible to prove that polyploidy occurs only when colchicine is present during mitosis.

These brief paragraphs can only suggest the hundreds of investigations that Eigsti and Dustin cite to better inform their readers about the role colchicine can play in the creation of polyploid plants and animals. In fact, the final chapter, "Mechanism of Colchicine-Mitosis," with its citation of 128 related investigations, emphasizes how "colchicine has become of world-wide importance and has opened new vistas in experimental agriculture."

To return to our earlier metaphor, no drama consists of only the current scene, much less of only the final act. If and when the final curtain falls, if and when life's myriad physicochemical features in times of both health and sickness are comprehended and become manageable, it will happen only through the coming together of the likes of such efforts as made the colchicine book possible.

O. J. Eigsti's primary interest in colchicine has always been its role in making seedless watermelons possible. It is to be expected that Dr. Dustin would be equally interested in the role colchicine has played and continues to play in the health sciences. Their book deserves to be recognized for its treatment of both roles.

There is a very special reason why colchicine has long held the attention of those who seek the origins of cancer and, ultimately, its prevention or cure. As pointed out earlier, colchicine can influence cell division, and the abnormally high rate at which cells grow and divide is cancer's most universal symptom. Quoting from chapter 10 of the colchicine book, "Neoplastic Growths—in Animals and Plants:"

> Mitotic changes induced by colchicine in a Crocker sarcoma of the mouse were described by Professor A. P. Dustin, Sr., in 1934. . . . At that time, the toll of life from bacterial diseases was declining as a result of the use of the sulfa drugs, and the relative incidence of cancer was gaining the impressive figure it has reached today in civilized countries. It is not surprising that the discovery of a specific action upon mitosis, the metaphase arrest, attracted wide attention. This research made clear for the first time the possibility of arresting cell division with chemicals acting specifically.

Thus advanced the method of cancer treatment, together with surgery, that dominated the medical profession throughout the twentieth century.

By reaching far, I find a shortcoming in the colchicine book, more a sin of omission than of commission, more often committed than avoided by those who seek explanations of nature's phenomena. The setting occurs in such statements as: "Cells of Alium with sixteen chromosomes as the diploid number accumulate chromosomes in hundreds, even more than a thousand per cell." Readers are left on their own to figure out how the requisite logistical problems are solved, how the necessary raw materials are obtained and the separate essential reaction mechanisms maintained for the replication of hundreds, if not thousands, of separate chemical syntheses in spaces no larger than a single cell.

In the physical sciences, the tenet that matter can be neither created nor destroyed is heavily emphasized. The atoms needed to assemble all of these new molecules must come from somewhere. Without taking these logistical essentials into account, the event as described above takes on the miraculous character of seven loaves and a few little fishes feeding the multitude. This may be a totally satisfying explanation, but it is certainly out of character when viewed from the predominantly materialistic approach to describing nature's events and circumstances.

Mitosis is one of nature's grand phenomena that introduce young people to life sciences. As such, it carries a tremendous pedagogical responsibility: It must be treated believably. If treated only superficially, we risk suffering from contradicting the "laws of learning" that tell us that "one learns mediocrity where one's talents are not exercised." If treated with no more than cold verbalization, we refute the "law" that "learning takes place only in activity. This learning may be mental, emotional, or physical, positive or negative, but to be actual, the learner has to participate." Furthermore, and in a sense that goes beyond the laws of learning, "education is far more than an intellectual process, it is vitally concerned with motives that underlie action. . . . Creation, growth, development, all depend upon stimulation and the rousing of emotional enforcement."

The public doesn't always appreciate the effort that goes into scholarly publications. The widely repeated motto "publish or perish" can easily discolor or pollute the true reasons why books such as *Colchicine – Agriculture, Medicine, Biology, and Chemistry* are written and published. Nor does this catchphrase suggest the spirit in which such books are written – the devotion to accuracy, honesty, integrity, the sense of responsibility that comes with reporting what other people have discovered and writing in a way that will help new investigators set out their pegs.

The role of colchicine in cancer research is a good example. Its effects on cell growth are its most evident property, and, again, abnormalities in cell growth is cancer's most evident symptom. It is axiomatic that each new attempt to describe the origins of cancer begins where prior research on the possible role of colchicine leaves off. Errors in reporting the results of prior research risk leading new investigations astray even at the risk of further delaying bringing the true origins of cancer into the open. On the positive side, accurate reports of prior investigations give new research efforts a head start. This is why Eigsti and Dustin took the writing of this book so seriously.

Again, from the point of view of making our lives count for something, it is curious how O. J. Eigsti has chosen two seemingly very diverse goals. One was to make high-quality seedless watermelons increasingly available at supermarkets and produce stands. The other was to bring together the results from hundreds of efforts to describe the properties of a substance and what these properties enable that substance to do. There is a world of difference in the satisfactions that derive from achievement of these diverse goals. Nor is it easy to establish the value of bringing together in one book the knowledge that is scattered among hundreds of sources.

There is no coin, no medium of exchange that can be used to put price tags on the gain from such human endeavors. The worth of what O. J. has done lies simply in knowing that he has been free to live life abundantly, pursue goals of his own choosing, make the most possible of his own resources, and help his colleagues do the same.

Something more comes to mind that is much like an afterthought. O. J. and his co-author cite some 1,300 men and women whose research has directly or indirectly involved colchicine. Other than T. D. Lysenko, none of the 55 members of the Lenin Academy of Agricultural Sciences in the Soviet Union who spoke at the previously cited sessions is mentioned, although many red pencil marks tell us that O. J. gave the published Proceedings much attention.

The general tone of these citations ruled out all possibilities that the use of colchicine would ever lead to improved sources of agricultural products. Furthermore, as discussed earlier with the exception of J. A. Rapoport of the Institute of Cytology, Histology and Embryology of the Academy of Sciences of the U.S.S.R., the members of the Lenin Academy would have us believe that just about everything reported in the colchicine book is ill advised at best, and very probably quite wrong.

Before I asked O. J. why he did not rise to this confrontation, I wondered if there was more than just an iron curtain separating two world powers engaging in the Cold War. Could it be that there was an intellectual curtain—something equally effective in separating the two nations and even more subtle, more fraught with risk for those who did not know how to keep their place or were seen as wanting to let intellectual light shine through? The timing here is significant. O. J. received his copy of the Proceedings in 1953. The colchicine book was published in 1955. The true nature of the fabric of either the iron or intellectual curtains had not yet been revealed.

The assumed intellectual curtain posed very serious risks, and there was no Berlin Wall to mark its presence. People could unknowingly brush against it in ways that enabled its unseen protectors to either welcome them as new comrades or punish them for exposing essential strategies. Although it would have had to be a different sort of book, Eigsti and Dustin missed a wonderful opportunity to document the ways and means and, ultimately, the results when powers outside the realm of true science turn its resources toward serving vested interests.

Chapter 7

From Seeds to Seedless

O. J. EIGSTI HAS said that his love affair with the seedless watermelon, his compulsion to make the growing and marketing of seedless watermelons a commercial success, began at that dinner party in Stockholm in 1948. The chance coming together of O. J. and Kihara was the first step along the quest that consumed much of O. J.'s time and energy. They met by chance but it would be misleading not to note that these two dinner guests brought with him his interest in colchicine.

For O. J., this substance had played the star role in his personal discoveries of its effects on cell growth, particularly on causing daughter cells to have twice the usual number of chromosomes. In Kihara's case, he knew how to use these effects to produce watermelons without seeds. O. J.'s interests traced to seeking to describe as accurately as possible a naturally occurring event; Kihara's interests were based on improving our food supply.

It was this chemistry at that dinner party that caused Kihara to visit the Eigstis on his way back home to Japan—a visit that led to the development of an informal joining of forces. Kihara would produce, in Japan, the seed that makes seedless watermelons. O. J. would promote the sale of this seed in the United States. At the same time, Kihara shared with O. J. his experience in growing and marketing seedless melons—experience that O. J. built on in his own research.

By 1954, all of the concerned parties were convinced that seedless watermelons would be a commercial success and the production of the essential seed would be financially rewarding. So, in 1954, O. J. organized the American Seedless Watermelon Seed Corporation and the Colchicine

Research Foundation. The $15,000 received from the sale of shares was to provide O. J.'s salary and cover other start-up expenses.

Things did not go well with the new corporation. Little profit was realized from the sale of the seed imported from Japan. Furthermore, O. J. was having trouble solving several seed production problems in his test plots. In 1956, the Polyploidy Crop Association of Japan made it possible for O. J. and Kihara to visit the major growers of seedless watermelons in Japan.

> *Kihara invites me to Japan as a guest of the Agriculture Department of Japan. The idea is that we are going to provide an outlet for the seed production in Japan. So the purpose of my visit is to see the seed production potential and make contact with the farmers with the help of the Extension Service of the University of Kyoto. Professor Kihara provided me with a marvelous opportunity to travel all over Japan. It was quite a trip. I was there at least six weeks from late July through August and into September. At this time, the farmers were harvesting their melons.*
>
> *They had this Polyploidy Crop Association and part of its work was to arrange a contest where the farmers brought in their finest melons. They were beautiful. The growers were very proud. The Japanese were almost more concerned about how they looked on the outside than the inside. They were round and they were beautiful! I was an honorary judge. We were to decide which was first, second, third, and so on on the basis of how they looked on the outside. Then they cut them and half of these melons were full of seeds. I mean they were full of seeds. They were not seedless. [The farmers] hung their heads and nobody knew what had happened. It was a disaster.*
>
> *We went to four different contests. Even before the contests, we visited the seed producers and I got to see everything that was going on in terms of seed production. They were having trouble just like we were back home, particularly in getting the seeds out. Finally they hired girls who would sit at tables just like at a picnic and they would pick out the seeds one by one. They'd cut the melons open and pick out the seeds. Then they washed and dried the seed.*
>
> *We also went to several markets where they were auctioning off watermelons wholesale. At this one Tokyo market they had melons piled up like a pyramid. And they were beautiful. But when samples were cut open and the auctioneer saw those seeds, he refused to sell them and as much as told the farmers to get rid of their melons. I came home knowing we couldn't continue to buy seed from Japan. Too much would not be seedless. So, what are we going to do? Actually, this was only one of the times I seriously considered giving up the whole business.*

O. J. didn't give up the watermelon business, but he did buy back from friends and relatives their shares in the American Seedless Watermelon Seed Corporation. He didn't like the feeling that he was causing them to lose money. More important, he turned the major chores of running the business over to his brother-in-law, who lived next door, and he and Agnes obtained full-time jobs in Chicago, O. J. at the Chicago Teachers College and Agnes at the General Biological Supply Company. Agnes made instructional microscope slides as head of the parasitology department and O. J. taught biological sciences. Weekends and summers they returned to their home at Goshen where they could stay in touch with the seedless watermelon project.

For the next twenty or more years, Eigsti's life was like trying to keep a computer operating four programs simultaneously. Until retiring in 1976, he prepared and presented his lectures at the Chicago Teachers College, later known as Chicago State University. He also traveled near and far beating the drum for the acceptance of seedless watermelons both at the supermarket and as a source of income for growers. This included solving problems, some downright disastrous, that no one could have foreseen. There was also the continuing effort to produce a seedless watermelon that would be the sweetest, tastiest, the very best melon, with or without seeds. And all of the time, O. J. did his best to stay up to date with genetics.

There were times when setbacks brought almost unbearable depths of despair. Finally, beginning in 1990, success soared to such heights that a single year's earnings more than paid back all the money that had been borrowed to keep the project alive. With this assurance that our story will have a happy ending, we now revisit events and circumstances along that tortuous and sometimes tortured trail that is the life of O. J. Eigsti.

Returning to meeting O. J. in that university greenhouse, I begin with his efforts to improve Kihara's tetraploid seed line. Traits that posed both minor and serious threats to financial success were hidden within the genomes of the tetraploids that Kihara had created. A minor defect was the hollow heart—rather than presenting a full measure of a watermelon's most favored portion, when these melons cracked open there was only empty space at the heart.

Susceptibility to disease is a very serious threat to the success of any variety of watermelons. This trait has an interesting history. For many years, it was easy to think of such clear-cut traits as eye color, hair color, and blood type as being controlled by genes. But something as vague as resistance to disease is another matter. Although first observed early in the 1800s, only recently has disease resistance been dealt with as an inheritable trait. Earlier, people were puzzled to see some watermelon plants sicken and die while their neighbors continued to flourish, both in their native habitat in the Kalahari Desert and in the new plantings in South Carolina. Somewhere, someone realized that there must be a reason. Eventually from this realization came the hypothesis that genes held the answer—a hypothesis that has been proven true. Interestingly, the watermelon seems to be the first species in which it was proved that disease susceptibility is an inheritable trait.

Furthermore, Dr. Taylor at the U. S. Department of Agriculture Experiment Station in Calhoun, Louisiana, developed a variety of watermelon, the Calhoun Gray, that was particularly disease resistant. O. J. tells how this variety helped him:

> *All these melons that Kihara produced, the tetraploids that [he] produced and the one that I produced, the 4008, were terribly susceptible to wilt. Before I went out to Riverside on my sabbatical, I sent to Dr. Paulus, a plant pathologist at the University of California at Riverside, some of my 313 seed line to plant in a test plot so when I came out I could see how they made out. The 313 variety was totally susceptible to wilt. Paulus really [raked] me over the coals. He said, "You should really be ashamed to send out seed that you know isn't going to grow." "Oh," I said, "what shall I do?" He said, "There's only one thing to do. Get Calhoun Gray, the diploid, and make a tetraploid." The Calhoun Gray is really wilt resistant. It's the standard. But it never caught on because it didn't have quality, it's sour. . . . So in 1976, I got seed from Taylor and I treated it with colchicine to make a tetraploid. You see, I realized that we would never go anywhere with seedless melons unless we had a disease-resistant variety. Fusarium wilt is a very serious disease. What would appear to be a healthy plant can be stricken and die overnight. Anthracnose is almost as deadly, particularly where the morning dew lingers. Downy mildew, gummy stem blight, and stem end rot pose less serious but always present risks.*

The actual genetic patterns for the inheritance of resistance to these diseases is not yet known, but there are ways to build disease resistance into a seed line. This is what O. J. is continuing to do when he keeps under close observation the 600 tetraploid melon plants in his test plots. The strategy is built on the assumption that one or more of these plants will inherit a full array of the genes that carry the disease resistance trait. This is something that will be sure to show up for a very special reason. The organisms that cause these diseases are soilborne, that is, they linger and multiply after flaring up in one generation of watermelons and O. J. knew that watermelons have been grown in this soil year after year for seventeen years.

Each of those 600 plants that survives and produces a mature watermelon presents O. J. with the equivalent of a lottery ticket, the kind where an opaque layer is scraped away to reveal what the buyer hopes will be the winning number.

And I know what I am looking for. I'm looking for a melon that is long, that weighs between twenty-two and twenty-five pounds, and has deep red flesh. Oh, you have never seen the red! It's a deep bloody red and if it was cut and put on the market even as a tetraploid with seeds it would just go to town. I saw that and I made a mistake, I made the mistake of not saving the seed and saying this is the selection, the melon we should work on. I had 60 melons in that test plot. I was getting awfully tired. I went through 60 different melons to get to that one melon. But I didn't save the seed. But I still have the seed line that was being tested in that plot. That seed produced the 600 plants in this year's test plot. That melon is sure to show up again and believe me, this time I will save the seed.

This seed line has an interesting history. I developed the Calhoun Gray tetraploid and was growing it up here in the garden. And while I was developing this tetraploid, Andrus at Charleston sent me wilt resistant Congo that he had developed. I planted it in the same garden and low and behold without my knowing it, these melons got mixed up, got hybridized by the bees, and the reason I know this is I picked some seeds out of a melon that was light rind and when I planted these seeds they produced melons that had dark rinds.

Actually, bees had done for me something I had tried to do but couldn't. One year I hand cross-pollinated 175 flowers and I think I got three sets. I gave that idea up and have relied on the bees for open pollination ever since. And it's much better to rely on open pollination 'cause I've got a theory here. The theory of gametic selection. The theory of pollen selection and if you open pollinate, if you let tetraploids open pollinate, selection will eliminate the less vigorous pollen grains and the male gametes they carry. And this is what is happening in these 600 plants in my test plot. If you hand pollinate, you bypass this opportunity to let nature get rid of inferior plants and with them the genes you don't want.

There's something more I want you to know about the seed that produced these 600 plants. When I was growing earlier generations here in the garden, I had trouble getting melons to ripen before frost killed the plants. So in 1977, I had Nick take plants down to Fort Worth in Texas where he was living at the time in September and had him finish them off. And that turned out to solve a very crucial problem.

In his test plot, O. J. stands to enjoy all of the satisfactions, the joys of success, the thrills that are to be enjoyed in other things that people do. Of course, he has known and will continue to know the disappointments and the unparalleled emotional and financial ups and downs that come when one plays games with nature.

O. J. was able to harvest the melons from these test plots before I wrote this book. Once more, he was amazed by how persistently unwanted genes expressed themselves. There were melons of all shapes and sizes with an equally wide range of rind colors and patterns. Melon after melon was cut open only to leave him with the same disappointment than comes with having paid for a losing lottery ticket. Then there was that one beautiful melon—a twenty-pound melon, oblong, with a nicely striped rind pattern, and with deep red and full-flavored flesh. Now the bad news. Somehow the seeds from this melon ended up mixed with the seeds harvested from less promising neighbors. Next year, O. J. will still be playing with chance, but he has increased the chances that his dream melon will be there at harvest time.

No matter what O. J.'s primary mission or responsibility, the seedless watermelon has always been, and still is, on his agenda. The stories he tells of his three years of sabbatical leave from Chicago State University provide good examples:

> *I returned to teaching in 1955 and seven years later, in 1963, I decided to take a freighter to South America. I was on this ship forty days and forty nights. We carried superphosphate, raw, in the hold of the ship from Tampa, Florida, to Brazil de Sol. How do you suppose they would get raw superphosphate out of a ship? By a bucket. Dust! Oh, it was awful. We were two weeks unloading that superphosphate. And I haven't been on a ship since!*
>
> *I landed finally in Buenos Aires, and this sabbatical was for travel and you were not supposed to be in any one country for more than a month. I had to send a card from a different country every month so I visited seven countries in South America. In addition to all of the things tourists are expected to do, I was looking into seed production, particularly in Paraguay. I wanted to visit the Mennonite colony in the Chaco to see if we could produce seed down there. This is 1963 and we still had not solved the seed production problem. We had that disaster in Florida in 1957 with the melons not being seedless and we had not solved that problem.*
>
> *The Russian Mennonites settled on this big plateau. This was the only country that would take them when hundreds were forced out of the Ukraine in 1931. It was wonderful watermelon country and there was the possibility that a South America[n] contact would let us grow two generations of experiments each year. But they had no way to get the seeds out of melons, so this didn't work out. The whole sabbatical was a wonderful experience and I fulfilled my travel obligations, but after bus trips back and forth across South America, I was ready to take a plane back to Miami.*
>
> *But I still had three months to go on my sabbatical. At Miami, I got on a bus and started traveling all over the U.S. I went to Charleston, South Carolina, to visit the U. S. Department of*

Agriculture's Experiment Station. In fact, except for returning to Goshen for Nick's wedding, I spent the three months visiting people in the watermelon business from Florida to Texas, California, and everywhere in between.

A letter from President Crispin Insaurralde at the National University of Asuncion, Paraguay, illustrates how O. J. arranged in advance to combine his roles as educator, scientist, and promoter of seedless watermelons. Dated 30 June 1962, the letter reads, "Being aware of your investigations and studies in the field of production of seeds and particularly in the development of seedless watermelons, we are pleased to invite you to conduct similar activities at the National University of Asuncion, College of Agronomy and Veterinary and Institute of Sciences. This office of Rector [President] believes that your knowledge on the production of seedless watermelons shall be very useful for our students in the Agricultural sciences."

Looking ahead to a one-semester sabbatical that would be due in 1970, O. J. lined up growers primarily in Texas, where he believed conditions were best for producing seed.

As soon as the first semester was over, we borrowed Nick's 1963 station wagon and headed for the Rio Grande Valley. At this time the focus was on Texas. People there are really becoming interested in seedless. So I got an appointment as a visiting research person at Texas A and M. I was writing then. I was working on updating my review of colchicine. I used the library to get together a lot of the literature on colchicine.

Through January and into March, I spent a lot of time visiting and studying the seedless situation. Not only in Texas but all of the way up to Oregon and Washington. I made many contacts. But I kept running into a dead end. Growers were interested, but they kept running into problems. Germination was a real problem. All of the time, we were trying to sell seed. But sales were very slow. They didn't amount to much.

But I met important people. There was Al Harrison, the "Watermelon King." His growers produce millions of watermelons each year. In 1949, he planted 20 acres with seed from Japan but was disappointed in the quality of these seedless melons. They had hollow-heart and it took years before he came back to growing seedless melons. Harrison is really quite a guy. One news article shows him riding astride a watermelon "airplane" and smoking a cigar that is larger than the plane.

On this trip I also met Bruce Price who had just left his job with the American Seed Trade Association and was taking over the family watermelon business. And Bill Burpee and I flew together to

Mexico City where we rented a car and drove all over western Mexico where watermelons were being grown. Bill Burpee was very excited about seedless watermelons, very excited. This is Bill Burpee the Fourth of the seed catalog family. He was very enthusiastic! When we got to Nogales, he took me to Tucson, where I took a plane back to Texas while he stayed to do more seed work. In December, Bill Burpee died of cancer. I am sure had this not happened, the whole situation with seedless watermelons would have taken a different turn. He had the resources behind him to really move this thing. That was a great loss, a terrible loss. What he was able to do was to write up the seedless melon story in a nice little brochure, which we made good use of. He did a really nice job.

On this sabbatical, this one for only one semester, Agnes went with me. On the others, she stayed with her work at General Biological, but by now she was retired. We drove this red pickup truck to California by way of Texas. But this time we concentrated on California. We got an apartment at Riverside, close to the University of California. During the day, I worked at the library but late in the afternoon we would drive around and visit watermelon growers in the fields where they were working. I remember being stopped one day by another truck. The driver thought we were federal agents looking for illegal workers. When he found out who we were he was greatly relieved. So were we.

We had a list of growers in the area. Some turned out to be antagonistic. But Harry Otto and Roy Kobayashi were very enthusiastic. For now, we came home all of the way through Canada still driving the red pickup. We went back to Chicago, since I had to fulfill my obligations by teaching one more year before retiring on August 30, 1976.

The papers presented at professional meetings have long been good opportunities for O. J. (There is more about this in chapter 11.) For now, I use a paper he presented in New Orleans on 1 March 1971 to illustrate his promotion of seedless watermelons. His audience was members of the National Watermelon Growers Association. The title, of course, was "The Seedless Watermelon."

He opens by regretting the small fraction of the public who have heard about this new fruit and its better flavor and finer texture than regular melons. He blames the size of this minority on unwillingness on the part of growers to admit that the seedless melon is here to stay. He assures his audience that the problems that have given growers trouble will soon be solved. There is no doubt but that consumers will soon be demanding seedless melons.

O. J. traces the development of the seedless melon both in Japan and at Michigan State University during the late 1930s. He compares its rate of acceptance by growers with that of hybrid corn and more than suggests that growers who hesitate to take up seedless melons will suffer losses as great as those suffered by corn growers whose conservatism caused them to stay with open-pollinated corn long after their competitors had shifted to hybrid. He pleads for help in raising the percentage of seedless melons being produced from 1 to at least 7 percent, an increase that could surely be achieved in a few years.

After briefly reviewing the science and technology underlying the production of seedless watermelons, O. J. assures his audience that the American Seedless Watermelon Seed Corporation appreciates the cooperation received from the seed trade and is ready to join in solving any problems that new growers may run into. In general, he remembers the audience's reaction as being not at all receptive.

There was another problem associated with O. J.'s early efforts to promote acceptance of the seedless watermelon. News reporters sometimes paid more attention to novelty than to getting their facts straight. One yellowed clipping with no identifiable source includes such statements as: "Dr. O. J. Eigsti, originator of American-bred seedless watermelons, was a young researcher for the Carnegie Institution of Washington, Long Island. He put his finger on the drug, colchicine, as a reliable means to induce the doubling of chromosomes in plants. Learning of the drug, Prof. H. Kihara of the National Institute of Genetics in Mishima, Japan, experimented with colchicine in the production of a watermelon seed which would yield a seedless melon. In the meantime, Dr. Eigsti confirmed in this country Dr. Kihara's suggested method of producing seedless watermelons."

Another example, equally yellowed, includes worthwhile information but garbles O. J.'s role in launching the seedless watermelon project. The caption for the accompanying photo reads: "Andrew Larson, Jr., of Havana, wearing a cap and glasses, throws a melon into a threshing machine to obtain seeds from it for other melons—which will be seedless. Dr. J. O. [sic] Eigsti of Chicago [wearing bib overalls] developed the method for producing this type of melon." The description of the threshing machine, something few people would recognize anywhere but in the melon patch, deserves to be quoted.

> Five years ago Larson started out raising five acres of the Tri-X variety melon and has increased his crop to 40 acres. He purchased a hand-built machine which is mounted on a wagon base with wheels. Powered by a gasoline motor—it grinds up the fruit, sends the rind out one end with the pulp and seeds pouring into a container.
>
> The mixture is put into a scrubbing machine with 40 pounds of water pressure. The pressure forces the pulp through a screen. The clean wet seeds are taken to a drying department where they are dried in forced hot air from 90 to 100 degrees.

> Larson harvests about 10 pounds of dry clean seed to an acre of melons. The melons that are used to produce the hybrid seeds have only one to two dozen seeds in them although they produce about one third more melons per acre than the ordinary melons . . .Usually he harvests the melons in the field, but found it easier to assemble all of the machinery in one location and haul the melons to that spot where they could be tossed into the machine.

When viewed in retrospect, the media's treatment of the development of seedless watermelons has been in O. J.'s favor. There were both positives and negatives but all of the time he was promoting the seedless watermelon, through all of his travels and speeches, his mind was equally on ensuring that the seed he produced would measure up to expectations such as appear in the current American Sun Melon brochure: "Tri-X313 Out Performs All Other Seedless Watermelons." Growers and consumers of this variety are assured that the hybrid vigor that is bred into this variety will adapt itself "to the fertilizer and water regimes for growing watermelons" and increased yield will result from "a longer harvest period than most seeded varieties." This "unique hybrid is adapted to short days of the winter-spring growing season, as well as the long days of summer growing. In fact, whether dry or wet, warm or hot, this hybrid is adapted to many climatic regions of the world." Finally, "with its firm texture, crispness of flesh, sweetness and extended shelf life, the Tri-X313 is superior to all other hybrids available to the trade."

This advertising copy strikes a thought-provoking note. Although its messages have been carefully crafted, they can only hint at the investment of resources, human and financial, scientific and agricultural, that make these melons possible. Many people have shared in this investment, contributing to and benefiting from its ups, being party to and suffering from its downs. All along the way are memories of things that worked out well, intermingled with regrets resulting from ill-advised decisions or unanticipated vagaries either in nature or in the market place.

A good example begins when O. J.'s duties as a geneticist with the Funk Brothers Seed Company took him to one of their seed plots in Mason City, Iowa. He tells the story:

> *I met the County Agricultural Agent and asked him who would be a good man to plant some seedless watermelons. And immediately he said, "Mr. Larson, Andrew Larson, Senior." So I went to see him. This is 1950, I had these seeds from Kihara. So here is this man, sixty-five years old, and he's just like a youngster. He said, "sure." That's wonderful! The average person at age sixty-five would say, seedless watermelon, this is crazy. And here is where I got started*

with the Larsons. This is very important because the Larsons planted that seed and then we had our first test plot. And I did all of this on the side while I was working for Funks.

If it hadn't been for the Larson family we wouldn't have survived in the watermelon business. In the sixties I was ready to [quit] and one day Senior Larson came to me and said, "I want you to promise me one thing." He said, "you're having trouble now but I want you to promise that you won't give this up. This watermelon has quality. Seedless is not the important thing. This triploid has quality. I've been growing watermelons all of my life." He's close to eighty. He said, "don't give up. Some day this is going to be worth something. Promise me you won't quit. I know you are having trouble and are discouraged now. But don't give up."

This is an example of the unforeseen problems that were behind Larson's reference to O. J. having troubles. The setting is a large farm in Monticello, Florida, where Stuart Simpson raises hundreds of acres of watermelons primarily for seed production.

I had met him at meetings of the American Seed Trade. I was going there because I was working for Funks on hybrid corn, but that didn't keep me from talking to watermelon people. And he wanted to produce seed for seedless melons. We had enough seed on hand to let him plant fifty-two acres of seed production, planting two rows of tetraploid and one row of pollinator.

So, several months later I got a call saying [I] better come down here. He said he had been going through the melon fields and was finding both seedless and seeded melons that had the same color and pattern rind. I told him I'd be right down. So I went to Monticello and I walked whole fields row by row, back and forth. Each time I came to a melon that was supposed to be seedless, I cut every melon on the vine and if one melon had seeds, every melon had seeds; if it was seedless, all melons on that vine were seedless.

Kihara had told me of a similar problem. He thought it was a matter of when flowers appear. Those that are first to appear might be seeded but, for some environmental reason, the later to appear flowers would produce seedless melons. Cutting all those melons told me that wasn't the right answer. The real cause was something else. It took years to solve this puzzle but now I know it was parthenogenesis—something that happened in the fields where the seed for this year's crop was produced. A few melons in that field grew on vines where one egg cell developed into a mature melon without being fertilized. The seeds in this melon would therefore have only two sets of chromosomes and be like ordinary diploid melons rather than seedless triploids or seeded tetraploids. And believe me, it took years to solve this problem!

Another example of discouraging circumstances grew out of the sale of 250 pounds of seed to another grower. When planted, not a single seed germinated—obviously a very expensive circumstance for everyone concerned. The reason turned out to be due to a difference between triploid and ordinary or tetraploid watermelon seeds. When ordinary seed is being harvested from ripe melons, the crushed flesh is allowed to ferment so the seed can be washed free. But triploid seeds, the ones that produce seedless melons, lack endosperm, the food storage portion of seeds. During fermentation, the empty space becomes filled with water and the embryo, in effect, drowns. This problem was not easy or inexpensive to solve. Seed harvesting machines had to be redesigned.

It was to be expected that seedless watermelons would become the topic for both professional journals and popular magazine articles. Hundreds of people began sharing the problems they encountered and how they solved them. The total of all of these efforts would be a fun story to tell, but it is almost impossible to give due credit to all the people who learned what had to be learned to make the whole story happen. There are dozens of little tricks of the trade that distinguish successful watermelon growers from those who fail—things learned from experience and shared more as friendly gestures than held back as trade secrets.

A good example of this kind of sharing is an eight-page photocopied "Progress Report" by J. Edward Klinker, distributed in 1953 by the Kentucky Agricultural Experiment Station. At that time, Klinker cited "two objections to the production of seedless watermelons. First, seed is expensive (from 2 to 5 cents per seed) and, second, because of a very hard seed coat, difficulty in obtaining satisfactory germination is often encountered." How Klinker set out to solve the first problem is a good example of finding and sharing solutions to problems.

> Trial packages of the seedless varieties Asashin were received from the Japan Seed and Plant Company, Tokyo, Japan, and the seedless variety Yamota [strain] No. 3 x Asahi from Fugita Seed Company, Ltd., Osaka, Japan. Instructions from these two seed companies included notes to maintain bed temperatures at 30°C. (86°F) after planting and either cut off the taper peaks of each seed or to soak in water. Actual techniques to maintain this bed temperature were not described.
>
> Seed of each variety was divided into two lots. Taper peaks of each seed of one lot of each variety were removed by means of a pruning shears while seeds of the other lots were left intact. Seed was then planted on May 19 in 2″ x 2 1/2″ tar paper plant bands arranged in ordinary wooden flats. Rate of seeding was one seed per band. Steam sterilized soil used in these bands was composed of two parts loam soil, one part rotted manure

> and one part sand. All flats were placed in the greenhouse on raised benches and thoroughly watered. Greenhouse day temperatures ranged as high as 95°F. and night temperatures as low as 70°F. Seedlings were transplanted to the field on June 1.

And the results of this experiment: "An average germination of 75 percent was obtained with all three varieties. There were no differences in percent of germination between seeds which had the taper peaks removed and those left intact." Not content with these results, Klinker explored a different hypothesis the following year. For this experiment,

> It seemed more logical to use only that seed that was commercially available to growers from American seedsmen and sold not as a particular variety but merely as Japanese seedless watermelons. Seed was purchased from Vaughan's Seed Company in Chicago.
>
> Since concentrated sulphyric [sic] acid and sodium hypochlorite (Chlorox) have been used to improve germination of certain seeds with very hard seed coats, these materials as well as tap water were used as seed treatments. Seven lots of 10 seeds each were in half-pint Mason jars. The following treatments were used: (1) non-treated dry seed (check), (2) 24-hour soak in tap water, (3) 48-hour soak in tap water, (4), (5), and (6) 1, 5 and 10-minute soak in concentrated sulfuric acid followed by a thorough rinsing in tap water, and (7) 24-hour soak in 1 percent Chlorox followed by a thorough rinsing in tap water.

When all of the treated and control seeds were planted and the rate of germination determined,

> From the data it appears that soaks of 24 to 48 hours in tap water were the most effective means of improving germination since these treatments produced 100 percent emergence. The non-treated dry seed showed 70 percent emergence which approached the 75 percent emergence of dry seed in the 1951 experiment. It is not understood why none of the seeds treated one minute with concentrated sulphyric acid emerged, while those treated for 5 and 10 minutes with concentrated sulphyric acid showed 90 and 30 percent emergence, respectively. Although the 5-minute sulphuric acid treatment showed 90 percent emergence, seedlings from this treatment as well as those soaked 10 minutes were weak and emerged three to four days later than those from the water treatments. Seedlings from Chlorox treated seeds showed 80 percent emergence, but they were definitely stunted.

A second and equally vital problem is brought to our attention when Klinker writes, "An extremely heavy mite population developed rather uniformly throughout the plantings within two weeks after vining started. This build-up in mite population occurred with such tremendous rapidity and intensity that considerable vine injury resulted before the mite control dusts could be applied. Once the dusts had been applied, however, excellent control was obtained. Mite counts made four days after dusting showed that 1 percent parathion dust gave 100 percent control of the above forms of mites, while 5 percent malathion gave 95 percent control of the active forms."

To see firsthand how seeds are coaxed to become seedlings and the seedlings nursed until they can survive on their own, join O. J. and Agnes Eigsti on a Sunday afternoon outing from their home in Normal, Illinois, to the Larsons' farm near the Illinois River. Along the way they will entertain us with stories of how things have changed, not only during their lives but as far back as 15,000 years—back in time when the tongues of a huge glacier finally ceased their relentless rebuilding of the countryside.

Almost every feature of the landscape is a souvenir left behind by glacial action. Lakes mark where the huge icy bulldozer dug exceptionally deep or its enormous weight depressed the preglacial soil. Piles of well-worn granite boulders were left stranded when the advancing ice no longer transported them. Then we come to features of the landscape that developed when the melting ice released billions of gallons of water to work its way down, eventually reaching the Gulf of Mexico. Perhaps some of that glacial water soaked into layers of deep, subsurface material to spawn the aquifers that now supply the many irrigation systems that appear as gangling beasts slowly inching back and forth across newly planted fields.

Of special interest to O. J. are the variations in soil types as we pass farm after farm. He knows that each crop, including, of course, watermelons, grows best in soil that meets its unique needs. And dozens of different types of soil exist in the Illinois counties that were covered by the glacier. Soil types depend on several factors beginning with the kinds of bedrock that may have been weathered to produce the original soil in that area. Winds transport soil from place to place; so do streams, especially during floods. Century-long cycles of growth and decay of vegetation add characteristics to the soil. Dark color soils developed where prairie grasses predominated; lighter soils developed under forests. Clearing forests, cultivating, fertilizing, draining, irrigating, in fact, all agricultural practices can change the character of the soil and, hence, its suitability for one crop or another.

O. J. and Agnes are quick to call attention to both good and bad judgment on the part of the farmers whose fields we pass by. A weedy field means that an herbicide was not applied properly or at the right time. Patches of stunted or yellowed plants are seen as evidence of fields not being properly drained. Fields where straight rows of uniformly vigorous plants reach from highway to horizon bear witness to farmers who know their profession and take pride in how well it is practiced.

At our destination, we find long rows of hotbeds, structures in which watermelon seedlings are protected from cold winds by transparent covers and their roots warmed by buried heating cables connected to electrical outlets. The beds where the seedlings are either weaker or stronger than their neighbors draw special attention. O. J. and the farmer try to account for special advantages or shortcomings in their location, the kind of soil that fills the hotbed, the amount and schedule of watering, and any other factors that might affect the vigor of watermelon seedlings.

When asked why three kinds of seedlings were being nursed in these hotbeds, O. J. drew attention to a flier that was prepared especially for the American Seedless Watermelon Seed Corporation's customers. Because this flier answers many other "from seed to supermarket" questions, its contents are summarized here.

The triploid hybrid watermelons are grown from seed that must be made each year. Triploid melons are sterile and do not produce pollen. Therefore, a regular diploid variety must be planted along with the triploids. And at the time of first flowers, bees should be placed in the field at the rate of two to three hives per hectare to carry the pollen from the diploids to the triploid hybrids. As for spacing, expedience has shown that highest yields were obtained when rows were five feet apart, and plants were four feet apart in the rows.

Soil temperature is critical in obtaining good germination of triploid seed. If direct seeded, soil temperature should be 70°F at a depth of four inches. When soil temperatures are below 70°F, it is highly recommended that the triploids be started in Jiffy-7 Peat Pellets. These small, net-enclosed, spaghum peat pellets contain adequate nutrients for the young seedlings. The planted Jiffy-7s should be kept at about 80°F for seventy-two hours. This is a critical period, especially for temperature, and light is not needed for germination. After germination, normal watering and daylight is needed. The seedlings should be ready to be set out in the field in about three weeks.

The seed should be planted "upside down" so the seedling will put its root down, pull up, and shed the seed coat more readily. The Jiffy-7s seem to solve the problem that if too much moisture is available during germination, the seed imbibes the moisture and the young sprout literally drowns within the seed coat. After germination, triploid seedlings seem to be more vigorous and grow faster than diploids.

The triploid variety TriX-313 will produce blocky, green-striped, thick-rind melons of eighteen to twenty-two pounds, with some melons going as high as thirty pounds. In these melons, sugar translocation from leaves to melons is made at night, and is best at night temperatures of 80°F or higher. Therefore, the warmer the nights, the sweeter the melon. In commercial production, it is not unusual to obtain 2,500 marketable melons per acre, since the triploids yield five picks as compared to the two picks expected for diploid pollinators in the same field. The best way to tell when a triploid hybrid melon is fully ripe is by the deep gold or orange coloring on the underside of the melons. Although triploids don't continue to ripen after being picked, these melons will hold up almost twice as long as diploids during shipping and storage.

The first few melons to ripen may contain small, "vestigial" seeds. This is a breeding problem and a genetic problem not yet solved. Many growers prefer not to sell the first "picks" as seedless so as to maintain high standards and protect their higher price. To summarize, hybrid seedless watermelons are a premium quality product. They require great care at planting, are difficult to pick ripe and to market properly. They have been extremely successful where all conditions of growth and marking have been met.

Conversation with the Eigstis while returning to Normal may have been influenced by our visit with the Larson family. They seemed to be representative of the traditional farm lifestyle. Their attitudes regarding many topics, their hopes and fears, their apparent philosophy was reminiscent of the times when I, too, was a "dirt" farmer. There was the same allegiance to nature, to the natural world that made possible the things they wanted and needed to do; the same realization that so many features of our lives were subject to the vagaries of nature and must be planned and adjusted to accordingly. The farm family's conversation suggested a strong sense of being in command of one's own destiny. Its members felt free to make their own decisions, their own mistakes, to enjoy the satisfactions that come with knowing they are producing what their fellow men need.

O. J. Eigsti's life has been complicated by sweeping changes not only in the milieu of his upbringing, but in the other domains of his professional career. It isn't easy to trace another's personal attributes to their origins. One's own experiences are sure to get in the way. It is fortunate that other authors in similar circumstances provide a somewhat generic infrastructure that protects the validity of interpretations based on individual experiences.

Passing by one abandoned homestead after another is sure to raise the question of how growing up on a family farm is expected to implant ethical values, moral standards, and other attributes of the human experience. Lauren Soth deals with this in his book *Farm Trouble* (1957). In a chapter titled "The Shibboleths," this widely recognized author writes about what it means to grow up on a family farm. This title is, in itself, disturbing. It suggests that he is dealing with something that is empty of real meaning. The validity of this suggestion, however, derives from allowing the genuine attributes of the family farm as it existed when Eigsti was growing up to be applied in the 1950s.

Soth tells us that "since colonial days—maybe since the days of the sturdy independent yeoman of the English Middle Ages—Anglo Saxon peoples have placed high social value on the family farm. In the early growth of democratic government, the independent smart landowner played an important part . . . He was the solid foundation of free suffrage, free speech, freedom of religion, and the other essentials of free government." But that was another age and "modern democracy cannot depend on the preservation of the family farm. In the United States only 13 percent of the people now live on farms of all kinds, and this percentage is dwindling rapidly." O. J. argues that Soth sees things wrong:

> *This man hasn't visited the good farmers around Morton in Illinois. What he should do is to visit those farmers. He'd get an entirely different idea of who the farmer is. You take my nephew Ron, they're a different breed. This idea that is in this book is totally wrong. That's because he hasn't been to see these people who are real farmers.*

In the O. J. Eigsti life story, it is this "dwindling percentage" that is most significant. It means that he has been required to interact more and more with people who have not shared his birthright, whose ethics and morals, ambitions, and ways of achieving them are alien to those of the traditional family farm. Out of necessity, he has adapted and adjusted, but never without recycling the lessons of another day. At the same time, there is one attribute of farm life, perhaps the most seminal of all, that has suffered no erosion. It is the urge to grow things, produce things, to have something material to show for what one has done, and something to look forward to. Far from being "empty of real meaning," this urge energizes each step along the way from seed to supermarket.

Returning now to another of the problems that O. J. has wrestled with for so many years, the goal is to produce seed that will always yield totally seedless melons. He believes that the commercial success of seedless watermelons depends on being able to assure both growers and consumers that when a melon that is expected to be seedless is cracked open, not a single fully developed black seed, not even an empty black seed coat, will appear. And the day may come when the small, white, vestigial remnants of seeds will no longer detract from the crimson color of the melon flesh.

When O. J. wrestles with this problem, he calls on just about everything he has learned about the hows and whys of plant growth. He has added many layers of precision to the descriptions of plant growth in his school science textbooks and all the way up the education ladder through graduate school lectures and seminars. He is deeply grateful to fellow scientists and horticulturists whose discoveries of new relationships gave him a boost along the way. His ultimate goal is to be able to enter the submicroscopic domain, where the contents of pollen grains mix with those of egg cells, and intervene not with the entire natural goings on, but with only those events that lead to fully developed seeds.

One wonders why O. J. caught the mystery of nature's most vital step up the evolutionary tree to the most advanced members of the plant kingdom. For all but a very few of the people who enroll in biology courses, nature's grand scheme of things, which culminates in plants that bear seeds, fails to survive in their minds beyond semester exams. O. J. is one of those very few. For reasons that exist only in the motivations that energize the teaching profession, Professor Witmer's presentation of the phenomenon of alternation of generations in that Goshen College botany course left an impression on O. J.'s mind that simply did not fade.

Few botanical phenomena run a greater risk of never being brought to life than that circumstance that freed the plant world from being dependent on spores and other purely vegetative ways to reproduce themselves. Bacteria, algae, lichens, mushrooms, mosses, and ferns soon became upstaged by new species that did more than simply set a new generation on its way. They literally packed a lunch and provided a protective coat that allowed the tender embryo to sleep through conditions that would be unfavorable for further growth. All of this is accompanied by embedding the seed in a structure that invites transportation, whereby each embryo need not compete for living space close to its beginning. Burrs that stick to passersby, "helicopter" blades that carry seeds whirling through space, and lacy parachutes work fine. But what could be more effective in getting seeds scattered far and wide than sweet, juicy, crimson flesh in a shiny green package?

For most of the world's people, it is enough to be able to consume the food that is stored in grains, fruits, and vegetables. For the few—O. J. Eigsti is a good example—there is the urge to better understand nature's ways and means, to lend a hand in making nature more efficient, better able to feed us. But when all is taken into account, even this very essential motivation may play second fiddle to the even more basic urge to understand nature, to feel at one with the natural world, to know and appreciate what it means to be alive.

Changing seeded watermelons to seedless boils down to retaining nature's ways to furnish seeds with a means of dispersal, but to bypass their full development. The secret lies in achieving the latter without preventing the former. The focus is on pollen grains and the tubes that allow their contents to travel to the flower's female domain. The trick is to divert the action of one of the pollen grain's sperm cells, but not interfere with the normal action of the other. Here we face an essential realization: We are completely at the mercy of nature.

Ordinary bees are as far as we can go in calling on outside help. They will bring the pollen grains to where the pollen tubes form, but nature's chemistry takes over at that point. It is up to us to examine each step along the way and find a vulnerable spot. O. J. knows that varieties differ in their tendency to leave vestigial seeds. This suggests that genes are involved, and we are back to the selection process as it is carried out in his test plots.

To date, O. J. has pretty much solved the problem of having fully developed black seeds occur in melons that are expected to be seedless. He has traced the cause to parthenogenesis; that is, embryos developing from egg cells that have not been fertilized. This state of affairs strikes at the very center of the seedless technology. Seedlessness depends on the chances of an embryo being produced when two sets of chromosomes in the egg cell produced by a tetraploid plant are confronted by a sperm cell containing one set of chromosomes from a diploid plant. The result produces a perfectly normal watermelon, but there is something about the coming together of three sets of chromosomes, the triploid condition, that does not allow seeds to develop. When parthenogenesis allows an egg cell with its two sets of chromosomes to form an embryo without fertilization, the resulting melon will have only two sets of chromosomes in its cells and thereby have the properties of an ordinary seeded melon. When it is tossed into the seed extraction machine at harvest time, its diploid seeds are sure to produce fully seeded melons in the next year's crop.

From the historical point of view, in 1939, Cheong-yin Wong, at what was than known as Michigan State College, used the parthenogenesis phenomenon to produce seedless watermelons. He did this by treating seedlings with plant hormones that were known to cause unfertilized egg cells to form fully developed fruit. Quoting his article in volume 89 of *Science*: "hormone treated watermelons were perfectly seedless but varied in fruit shape. . . . Some, however, were normal in shape and size. The texture of these fruits was very solid and firm. No differences in flavor could be detected from normally pollinated fruits."

It is interesting that Wong also experimented with colchicine. The seed of three plants of the Winter Sweet watermelon had been subjected to colchicine treatment before planting. The plants showed a typical colchicine effect, i.e., stunting early in the season, large size of leaves and flowers and great vigor later in the season. Although pollen was present in great abundance, it failed to induce fruit setting when the blossoms were selfed. On the other hand, hormone-treated flowers set very satisfactorily. Fruits were formed in some open-pollinated flowers (with plenty of seeds), probably due to fertilization from nearby normal plants (vicinism).

It is not clear from Wong's writing whether or not these fruits were triploids. If they were, and had he planted their seeds near those from ordinary diploid melons, entirely new paragraphs would have to be added to the seedless watermelon story.

Chapter 8
From Being Educated to Being an Educator

WHO O. J. EIGSTI is, what he stands for, and what he has achieved has been greatly influenced by the teaching profession as a consumer and a producer, as a student, and as a university professor, The how and why of this influence is difficult to document for many reasons. A person's functional education has been likened to the few grams of residue left on the final filter, the sought after precious substance after tons of raw source material have been extracted by dozens of solvents and the inactive portions allowed to trickle down the drain. This analogy, however, may be somewhat misleading. The magnitude, the character of that final residue may well depend on the actions of elements totally essential along the way, but rightly abandoned after their role had played out.

In no way is it possible to sort through the "tons of raw source material" that yielded the final residue that is O. J.'s education. Nor can I fully document how he acquired the knowledge and know-how that made him a successful teacher. The best I can do is to review education-related events and circumstances that may have left their mark on him.

I begin with School District Number 47's traditional one-room country school near Morton, Illinois. Today, the image of a one-room country school house tends to be clouded more with nostalgia than with reality. What it was like to be one among thirty or so other students, some younger, others much older, with only a minimum amount of space, is easily distorted by romanticism. Only those who were there can know what it is like to take one's turn to sit on the front row to "recite." (Recitation could be a valuable learning experience for everyone in the room, but always at the risk of hurt feelings when shortcomings were displayed for all to see.)

Teachers and peers share in being remembered. When O. J. recalls his grade school teachers, I've heard such phrases as "strict in discipline but short on educational techniques." Their capabilities came more from dedication and devotion than from formal pedagogical training. In some cases, a high school diploma, supplemented by a summer course in pedagogy, was all that separated the teacher intellectually from her eighth grade students. It is to be noted, however, that such names as the Papenhaus sisters, Imogene Ford, and Emma Moser come easily to O. J.'s mind these many decades after they and the one-room Maple Grove school exist only in memory. As for his peers. one is cited most often:

> *Mary Bernhart was a brilliant girl. She was in the eighth grade with me. She came to school in the first grade and couldn't speak English. She was bright. She was a competitor. She and I competed. Very much so. In spelling and what not. She was a stimulus, an excellent student. And her father wouldn't let her go to high school. That was a terrible thing. He was an old German, conservative. He didn't believe in education.*

If we read between the lines of this one quotation, we grasp much of what it really means to be a product of the educational process as it operated on the American frontier. The psychological and sociological currents ran deep. The students' motivations came more from themselves and their sociological environment than from textbooks on educational psychology or counseling. Competiton with peers was fraught with fantastic highs and equally devastating lows. Even at the high school level, sometimes students in the more specialized courses where enrollment was low would "recite" in the front of the large study hall. That was an additional responsibility of the teacher. And tbere may have been times when those in the front row were chided or complimented with special emphasis, more to influence the inadvertent audience than for more honest reasons.

Viewed from the perspective of todays educational principles and dictates, the psychological implications of the "one-room" approach to schooling could well be seen as harsh, even to the point of being less than humane. In contemplating how this affected O. J., we must factor in a sociological demand that was very much a part of his growing up in a prairie state. Odds were against all sons and daughters finding the wherewithal to build their lives in the communities where they were bom and began their education. In the quotation where O. J. remembers his special peer, Mary Bernhart, there is more than empathy with someone whose family's nation of birth had so recently been a mortal enemy. There's more:

Actually, we lived in a community that you might call anti-intellectual. We lived in a community there in Morton where high school was something you should avoid. See, the Amish, the new Amish, that is, wouldn't send their kids to school beyond the eighth grade. The whole community was not intellectual. We didn't have any culture. There was nothing like music or lectures or anything like that. In high school I didn't study very much. I went through high school without really studying. I was able to pass tests without opening a book. I never worked. I never studied. I didn't really want to go to high school. I wasn't interested in school But I was a good student, I had the intellect, the capacity to be an outstanding student. I was the highest standing boy in the class and Anna Gest was the top girl.

We had no biology in high school, only physics. They didn't have biology at Morton High School. From 1925 to 1926 there were only four teachers. No biology, no chemistry, just physics, physics and math. They had a physics teacher who also taught manual training. My high school education was poor, very poor. We played basketball and track and baseball.

My motivatian didn't start until I got to Goshen College. Well, actually, it started at Bradley when I hit this psychology teacher. That's when I really started to study. When I saw it was so easy to get a B, why not just go ahead and get As. So I started studying to get good grades. And I started studying. I was a major in history, social science. But I took a chemistry course at Bradley. I enjoyed chemistry.

At this point, O. J. was reminded that the general science textbook his sister studied in high school had things to say about such topics as protoplasm and the theory of evolution.

Protoplasm! No, I had never heard of protoplasm. The word protoplasm never came into my view. Not until I got to Goshen College. See, when I started taking botany from Professor Witmer, that's when I really saw a new area of interest, I took every course he gave; botany, zoology, bateriology, genetics. I took my first course in genetics from him and I saw this Mendel situation and I was able to figure out these ratios and stuff and the other students in the class couldn't understand how Mendel worked and we'd get in the dorm and I would explain to them how Mendel worked, how this three to one and nine three three to one worked. I was really a student teacher.

As O. J. moved up the education ladder, he had to adjust to marked changes in how textbook authors treated genetics. At the same time, the contents of these textbooks became amazingly valuable to O. J. Eigsti as is well illustrated by George W. Hunter's *Civic Biology* published in 1914 by Macmillan. Citations from this widely used textbook follow:

"HEREDITY AND WHAT IT MEANS" This wonderful force which causes the likeness of the child to its parents and to their parents we call heredity. . .If we trace the workings of heredity in our own individual case, we will probably find that we are molded like our ancestors not only in physical characteristics but in mental qualities as well. The ability to play the piano or to paint is probably as much a case of inheritance as the color of our eyes or the shape of our nose. We are a complex of physical and mental characters, received in part from all our ancestors.

"VARIATION" Heredity and variation are the cornerstones on which all of the work in the improvement of plants and animals, including man himself, are built.

"The Bearers of Heredity" We have seen that somewhere in every living cell is a structure known as a nucleus. In this nucleus, which is a part of the living matter of the cell, are certain very minute structures always present, known as chromosomes. These chromosomes (so called because they take up color when stained) are believed to be the structures which contain the determiners of the qualities which may be passed from parent plant to offspring or from animal to animal; in other words, the qualities that are inheritable.

"THE GERM CELLS" But it has been found that certain cells of the body, the egg and sperm cells, before uniting contain only half as many chromosomes as do the body cells. In preparing for the process of fertilization, half of these elements have been eliminated, so that when the egg and sperm cell are united they will have the full number of chromosomes that the other cells have.

"CHARLES DARWIN AND NATURAL SELECTION" The great Englishman Charles Darwin was one of the first scientists to realize how the great force of heredity applied to the development or evolution of plants and animals. He knew that although animals and plants were like their ancestors, they also tended to vary. In nature, the variations which best fitted a plant or animal for life in its own environment were the ones which were handed down because those which were not fitted for life in that particular environment would die. Thus nature seized upon favorable variations and after a time, as the descendants of each of these individuals also tended to vary, a new species of plant or animal, fitted for the place it had to live in, would be gradually evolved.

"MUTATIONS" Recently a new method of variation has been discovered by a Dutch naturalist, named Hugo de Vries. He found that new species of plants and animals arise suddenly by "mutations" or steps. This means that new species instead of arising from very slight variations, continuing during long periods of years (as Darwin believed), might arise very suddenly as a very great variation which would at once breed true.

"ARTIFICIAL SELECTION" Darwin reasoned that if nature seized upon favorable variants, then man by selecting the variations he wanted, could form new varieties of plants or animals much more quickly than nature. And so today plant or animal breeders select the forms having the characters they wish to perpetuate and breed them together. This method used by plant and animal breeders is known as selection.

"SELECTIVE PLANTING" By selective planting we mean choosing the best plants and planting the seed from these plants with a view of improving the yield . . . The work of Mendel when applied to plant breeding will greatly shorten the time required to produce better plants of a given kind . . .

"HYBRIDIZING" . . . If pollen from one plant be placed on the pistil of another of an allied species or variety, fertilization may take place and new plants be eventually produced from the seeds. This process is known as hybridizing and the plants produced by this process known as hybrids. Hybrids are extremely variable, rarely breed from seeds, and often are apparently quite unlike either parent plant. They must be grown for several years, and all plants that do not resemble the desired variety must be killed off, if we expect to produce a hybrid that will breed more plants like itself. Luther Burbank, the great hybridizer of California, destroys tens of thousands of plants in order to get one or two with the characteristics he wishes to preserve.

In addition to changes in factual content, O. J. has had to adjust to changes in how authors treat possible applications of genetic-related concepts. An illustration of this begins with excerpts from *Elements of General Science with Experiements* by Otis Williams Caldwell and William Lewis Eikenberry and published by Ginn and Company. This book was published in 1926, the year O. J. spent on the family farm after high school and before continuing his education. Representative sentences include:

> We cannot emphasize too strongly the importance of making sure that no avoidable tendency to disease and no abnormally weakened bodies are passed on to succeeding generations . . . It is very important that the human race should recognize and use what is known of biological inheritance . . . It is apparently easier for men to see and appreciate the advantages from better cattle and pigs than to see the advantages in improving the quality of the succeeding generations of the human race.

In 1928, the American Book Company published Bertha M. Clark's *New Introduction to Science.* The chapter, "Imposing Our Will On Plants," assures us that its contents would relate to what became O. J. Eigsti's chief mission in life. The book is cited here, however, only for the author's rebuttal of points of view stated by the previously cited authors: "Remember always that what we are (heredity) and what we have (environment) are not our masters. They are our tools. and it is what we do with our tools that determines what we become in life."

The beginnings of O. J. Eigsti's thirty-five-year teaching career are wonderfully representative of the circumstances that brought new people into the profession during much of the twentieth century. When he followed his brother, Mahlon, to Bradley University in 1927, it was the faculty in the social science department who most effectively nutured O. J.'s intellectual interests. Although most names escape O. J., Professor Wychoff, the history teacher, and the general psychology and the education psychology instructors stand out in his memory.

When O. J. transferred from Bradley to Goshen College, he encountered what he claims was positively the worst teacher in his entire twenty years as a student. If we read between the lines, the irony of this memory is that this "worst" teacher may very well have contributed much to O. J.'s own performance as an educator. For example, he always gave his students a fair exam and graded them fairly. Often he would invite them to meet with him at the end of the semester and ask them what they thought their grade should be.

Less ironic, but probably equally representative of how things happened on college campuses at the time, it was at Goshen that a very inspiring teacher sent O. J. on his way to become a geneticist. When he could not get a teaching job in 1931, he went to the University of Illinois and began studies in botany. Here O. J. tells it:

> *Remember this is 1931. No job and the Depression is building throughout the USA. Farm prices are low, many have lost their farms. This is where the Depression really started and the worst is yet to come. So I decide to go to the University of Illinois to study the plant sciences. My experiences at Goshen College had finished me on education and psychology, but I was really excited about botany. When I got to Illinois, I learned that Goshen College was not accredited so all I could do was to enter as a student without standing. I signed up for the basic courses in botany.*
>
> *I soon ran into bad luck. The first weekend of October I hitch-hiked home for a visit. I was paying six percent interest on the money I borrowed so, no bus ticket for me. There I came down with a bad case of appendicitis, which sent me to the hospital. When I got back to the University three weeks later I was surprised to find that my instructors had missed me. They were sympathetic. In fact, Professor Buchholz said*

here is the lab, do whatever you can but he seemed to think I had no chance to get ready for the semester finals. But I did and when I got As in all of my courses, the instructors took another look at me. Someone, it may have been Dr. Stella Hague, a specialist in mosses and liverworts, had my status changed from non-graduate to graduate. She was the only woman in the department and knew what it was like to be at a disadvantage. Also, I knew where to find all of the local mosses and liverworts.

The next year I had an assistantship in botany that paid $600. This let me finish the work and receive my Master's degree. I spent the summer with Professor Buchholz at Cold Spring Harbor where he was doing genetics research. I skill expected to become a high school teacher, but when I couldn't find a job, I went back to work for the Ph.D. under Dr. Buchholz at Illinois.

Only those who have had the same experiences can fully appreciate what O. J. achieved in order to earn that Ph.D. The final hurdles were examinations based on the contents of a dozen or more courses and the final oral defense of a dissertation based on research of his own design and execution. It is to be particularly noted that his examiners were survivors of this same ordeal. They had faith in its purpose and were not about to allow its integrity to erode.

The bookshelves in the previously mentioned abandoned turkey barn held a textbook that gives us a glimpse of the course content that O. J. was expected to master. *Plant Physiology* by Edwin C. Miller and published in 1931 by McGraw-Hill is a 900 page summation of just about everything that was known at the time about what goes on inside the plant world. Each of the fourteen chapters includes citations of the 200 or so investigations that made the book possible. To give instructors a hand, 679 questions appear at the end of each chapter. For most of us it is enough to know that papain will tenderize meat, but students of this textbook are expected to answer question 580, "What is the reaction of papain to its substrate at temperatures from 80 to 90°C? Why? What proof is there for your answer? Describe a method for determining the action of papain upon peptone. Upon uncoagulated egg white. Upon milk" And papain is only one of the thousand or so substances that appear in the index of this book!

O. J.'s masters and doctoral theses were major milestones in his education. (They also served to gain membership in the science enterprise and, therefore, are discussed in Chapter 11.) Scarcely two years after receiving his doctorate, necessity became the determining reason for O. J. to resume his interest in becoming a teacher. Virtually unemployed and facing the bleak economic conditions of the Great Depression, he considered himself to be very fortunate to obtain a one-year appointment as director of physical education, coach of men's athletics, and professor of biology at Greenville College in Illinois. This was ten years after he started college with the goal of becoming a teacher.

Both the employment interview and the dimensions of his duties seem to be unbelievable but, they tell us where both O. J. and the teaching profession have been during the twentieth century. The interview was held at the Polk Station in Chicago. He was asked if he smoked, went to movie theaters, or danced. When Agnes was asked if she would remove her wedding ring, the answer was no, but it was decided that this could be tolerated. (Agnes remembers that no sooner had she gotten to Greenville, than a skin rash required removing the wedding ring.)

O. J. learned that the students at Greenville College were not allowed to smoke or go to the movies; if they went to the beach, they were only to go into the water, come out, and leave the beach. It is to be noted, however, that all seven of O. J.'s premed students went on to become successful physicians. O. J. doesn't say how the sports teams he coached made out, but he does tell us that he and Agnes were chosen to be the junior class faculty sponsors.

This appointment was for only one year, so O. J. continued looking for a more permanent position. This included asking for help from Professor Buchholz, his doctoral advisor at the University of Illinois. Before the school year was over, he received two job offers, from which he chose the position of assistant professor of botany at the University of Oklahoma. During the intervening summer, O. J. held a position at Goshen College teaching human physiology, zoology, and bacteriology. This became one of the many experiences the Eigstis shared with Goshen College.

This summer also serves well to illustrate the somewhat catch-as-catch-can manner in which the Eigstis have solved the everyday problems of living. While O. J. was teaching at Goshen College, relatives, Dr. and Mrs. F. S. Martin, provided a home for them. Agnes was pregnant and was willing to cook for the extended family while Mrs. Martin worked in her husband's office. Agnes also worked at the local hospital, which ensured tender, loving care for the impending birth. In fact, their son Karl arrived at four o'clock on the same morning that O. J. gave his first lecture at Oklahoma.

Geography and theology combine in many ways to give Goshen College a major role in the lives of O. J. and Agnes Eigsti—a role that speaks to relationships between institutions of higher learning and their students. How these relationships reappear decades later in attitudes toward each other is also beautifully illustrated by reviewing the nearly seven decades during which O. J. has known and been known by Goshen College.

These relationhips began in a spirit of conflict when O, J. transferred from Bradley University to Goshen College against his father's advice. The reasons for doing so were complex. At that time, Goshen was facing a bleak future. Conflicts within the Mennonite hierarchy were influencing the college's administrative function at the expense of student enrollment. O. J. responded to being recruited to come to Goshen. While at Bradley, he achieved athletic recognition. This led to being offered a position at Goshen as assistant athletic director, a position that paid his tuition and added to his stature as a student. But it all came with disadvantages. O J. had serious trouble with the rule imposed by the college's bishops that stated that male athletes had to wear shorts that were no less than twenty inches from the knee to the bottom edge of the shorts. Memories such as this compete with the happy memories that came with setting track records in the running broad jump, the high jump, pole vault, 100-yard dash, and the 220-yard dash. There were also academic issues.

> *At Bradley, I had very good teachers in the social sciences and education. Soon after I came to Goshen College, to continue in education was the last thing I wanted to do. But a very good teacher in botany caused me to turn from education to botany. And when I went to the University of Illinois they would not accept my credits in botany because Goshen was not accredited. As things turned out, however, what I learned in Professor Whitmer's courses really gave me a leg up in my graduate work at the U of I.*

With Goshen College's roots being deeply planted in the Mennonite religion, it is to be expected that conflicts within the church would spill over and spawn conflicts in both the college and the Goshen community. Such conflicts certainly involved the Eigstis, but it seems appropriate to say that such matters affected no more than the fringes of their lives. Hassles and turf battes within offcialdom may have left their scars, but one has the feeling that both O. J. and Agnes remain true to the best of what Mennonites believe in, and look on the Goshen College of today accordingly.

Agnes and their new son soon joined O. J. at the University of Oklahoma, but other things were left behind at both Goshen and Greenville colleges—things that distinguish the church-related concept of higher education as it was conceived in the early half of the century from its counterpart in colleges and universities that claimed no ties with any organized religion. O. J. continued to feel free to teach the subject matter of his courses as it had been taught to him. He expected and received the cooperation from his students that he was accustomed to. The big difference was in the interests and responsibilities of the men and women who held administrative positions. In his new position, O. J. believed they were interested first and foremost in the intellectual development of their students and left other interests pretty much up to each student.

During the years between 1938 and 1945, O. J. advanced to the rank of full professor at the University of Oklahoma The changes in what it means to be a college professor that evolved dunng these years are very significant. Initially, O. J. did the things he learned from his professors while he was earning his doctorate. His primary writing project has already been mentioned. He also carried out botanical investigations and reported their results in science literature. What was pretty much a new role for university professors began rather casually in 1938 when the Eigsti family had renturned to Morton for a visit with relatives.

> *My brother Emery was a salesman for Funk Brothers Seed Company for southeastern Iowa. He said he was going to Bloomington for a company meeting and asked if I wanted to go along. So I went with him. And at this meeting, one of the company executives was talking and I raised a question that seemed to strike at the center of something vital to the company. I didn't realize that I was raising a very sensitive issue. But I soon found out it was. It involved where the company had obtained the seed lines that made their whole business of producing hybrid corn seed possible.*
>
> *After the meeting the company executive asked my brother who was that young fellow. So Emery introduced me to him. Two years later I got a phone call at Oklahoma telling me that theFunk company wanted some test plots planted in Oklahoma. Could I do it? I said sure. I did it mostly on weekends and Saturdays. I went around close to Norman and set up research test plots with farmers. This went on wntil about 1943, when the head of my department said if you can get research funds from Funk—our enrollment is dropping so bad with so many of America's young people being drafted—we will be able to support you and your work. So I was the founder of the Oklahoma Research Institute. Funk Brothers sent the money to the university and they paid me out of that fund. I worked there until 1945.*

All of these maneuverings seemed to work out to everyone's satisfaction. As a college professor, O. J. met whatever classes needed to be covered. His work as a consultant added to his image among his students and colleagues. An undated yellowed newspaper clipping shows O. J. costumed somewhere between that of a college professor and a farm hand—the tie is the giveaway—as he stands between cornstalks taller than he is, with husks turned back to reveal healthy ears ready for harvest. O. J. holds all of the ears one arm can manage. The caption reads, "Progeny number 11 x 2-1507, one of the outstanding progenies found in inbred 2-1507 progeny testing program. Progeny 2-1507 had a very good growth record, both as a line and in hybrid combination during the season. It also had a good record in each location in which it was tested. At harvest this progeny was selected for ear and kernel type in hybrid combination with inbred 11, producing an ear highly desirable as seed parent to prcduce more than average percentage of regular flats." As O. J. remembers it:

> *That became the leading hybrid corn for the Southwest. It is still used today. I put together the inbreds based on their preformance. In other words, I put together this hybrid and tested it and showed that this is a good adapted hybrid for the growing conditions of the southwest. And that is typical of how I got involved with the Funk Brothers Seed Company.*

To see O. J. standing alone in a cornfield, it is difficult to associate the dimensions of what he was doing in the early 1940s with the large scale research agencies that today, fifty years later, owe their existence to affiliation with a university. The enormous laboratories associated with such universities as Johns Hopkins, Stanford, and California, with dozens of highly trained scientists joining forces in research teams and equally capable people maintaing adequate funding, had to have their conception somewhere. Wherever and whenever this happened, it was surely paralleled by the ultimate outcome of Emery's casual introduction of O. J. to the Funk Brothers Seed Company.

O. J.'s statement, "So I was the founder of the Oklahoma Research Institute," needs to be fleshed out. Which I can do with the help of *The University of Oklahoma Research Institute 1941-1973*, a book written by George Lynn Cross in 1986 and published by the university. For perspective, we begin with a citation attributed to Victor Barnett of the Tulsa Tribune, when he addressed a group of Tulsans sometime during the early 1940s. Barnett said: "The Oklahoma farmer used to find life a little confusing. He would get up at the alarm of a Connecticut clock, button his Chicago suspenders to Detroit overalls, wash his face with Cincinnati soap in a Pennsylvania pan, sit down to a Grand Rapids table and eat a Battle Creek cereal, Chicago meat and Tennessee flour cooked with Kansas lard on a St. Louis stove. Then he would put a New York bridle on a Mississippi mule, fed with Iowa corn and plow a farm covered with an Ohio mortgage with a Moline plow. At the end of the day he would say a prayer written in Jerusalem, crawl under a blanket made in New Hampshire, only to be kept awake by an Oklahoma dog—the only home product on his place. But, hopefully, it may be gettng a bit different now."

The University of Oklahoma Research Institute was officially established when the articles of incorporation were filed with the secretary of state of Oklahoma on March 29, 1941. This followed the development of several research projects, most of which involved petroleum. However, quoting Cross,

> One project, rather unusual for the University of Oklahoma because it involved agricultural research, was

sponsored by the Funk Seed Company and involved comprehensive experiments with several strains of hybrid corn. The project was directed by Orie J. Eigsti, of the Department of Botany. The objective was to produce a new strain of hybrid corn that would mature into at least a fair crop in the Southwest regardless of weather conditions.

Considering the ultimate impact the oil industry had on Oklahoma, it could well be that O. J.'s work on hybrid corn did not loom large in establishing the circumstances that gave birth to the Oklahoma Research Institute and nutured it through its infancy. But O. J. was there when it happened. Futhermore, we know that many Oklahomians would feel more positive toward reasearch if they could see the fields of corn ripening and ready for harvest well before an early frost could wreak havoc thereby having something more than a barking dog to show for Oklahoma's productivity.

O. J. further explains his situation in 1945:

> *I wasn't hired by Funks. I was a consultant. Actually, I was spending so much time with the Funks that we bought a home in Normal even though my official place of employment continued to be the University of Oklahoma. By 1945 it made better sense to take a job in the biology department of Northwestern University in Evanston. But I continued with the Funks. I found farmers near Evanston who carried on the testing for the Funks.*

When O. J. joined the faculty at Northwestern, he found a university caught up in changes that were sweeping across academia. The GI bill brought a flood of students where previously there had been an extreme shortage. And these students were a new breed. Many had been practicing what their instructors could only talk about. The values, ethics and morals that had enabled them to survive during the past few years were a far cry from those that were reflected in the pre-employment interviews at Greenville or Goshen College. But GI students adjusted with amazing finesse.

As was the case with many universities, Northwestern's administration saw the arrival of this flood of new students, with their tuition guaranteed and their expectations high, as an opportunity to expand, upgrade, and explore new frontiers. For some universities, utilization of this opportunity yielded fantastic results. "Cow Colleges" became "Big Ten" universities. Regional liberal arts colleges earned national recognition. And Northwestern set its sights on finding its place among the nation's most progressive and research-oriented universities.

O. J.'s four years at Northwestern between 1945 and 1949 provide an opportunity to examine how and why some faculty members are comfortable with and benefit from their administration's ambitions while others suffer. O. J. advanced to the position of tenured full professor. Agnes's memories are those of an exemplary "faculty wife."

> I taught one full year [nutrition] at the same time I was running O. J.'s lab in our basement, was the social chairman at Karl and Nick's PTA. Had the title of some sort on the Welcoming Committee of New Faculty who were moving into the Quonsets, and on a committee at the First Methodist Church's Women's Society. We had an Assyrian housekeeper, whose daughter worked for me in the lab, plus the wife of a young ministerial student. And there was that faculty dinner party with twenty guests, a whole turkey at one end of the table and a Morton ham with O. J.'s initials carved in the rind at the other, scalloped oysters, homemade rolls, etc., and a bar in the basement.

O. J. recalls one specific point of contention. The botany courses he taught were to be blended into more general life science courses. He saw this as a threat to the academic standards that marked his advance to the doctoral level and gave him license to call himself a professor. The threat may have been real, but it probably wasn't enough to cause O. J. to abandon, at least temporarily, his role as a university professor. Usually, under such circumstances, there are subtle and more serious reasons for such action. And although this may not have been the case at Northwestern, it is notoriously easy for small-minded and self-serving members of the traditional "appointments, promotion, and tenure" committee to work their way into positions of much authority. In any case, in 1949 the Eigstis left Northwestern and moved back to Normal. O. J. devoted his time to the Funks consultantship position but, of course, he found time to continue the development of the seedless watermelon.

In 1952, O. J. was appointed Fulbright professor and visiting professor of genetics, University of Dacca, East Pakistan, and the University of Punjab, West Pakistan. This story begins with casual conversations with friends while visiting the University of Maryland and includes without doubt the most exotic of all variations from the life the Eigstis had known. When the two airline tickets arrived from the State Department, the Eigstis realized they could choose between east and west for what could be a trip around the world. After making the arrangements required for being out of the United States for at least a year, including a home and "parentage" for sons Nick and Karl, they chose the westward option.

That the Eigstis would be half a world away from family and friends, things familiar and a comfortable lifestyle, tells us as much about Agnes as about O. J.. The biblical, "Whither thou goest, I will go," comes first to mind. Somehow that doesn't fit Agnes, it is too passive, too suggestive of subservience. "Tell me where you want to go, I will help you get there," seems to fit better. Here is an example. On the way to the Chicago airport,

Agnes carried a larger than usual purse. It was stuffed almost to the bursting point. Her explanation, "We had been told that the American one dollar bill would solve just about every problem one runs into while traveling, especially in the Mideast." And Agnes was ready to solve lots of problems.

> *The day before we left, a few of our friends at Normal had a farewell luncheon. This was much appreciated. We visited friends in California and agian after arriving in Hawaii. A typhoon knocked out facilities at Midway, so we were delayed on the flight to Japan. We finally got to Japan and had a good time with Kihara and his associates. We also visited friends at the International Christian University. Stopped in Hong Kong, then Bangkok for a few days, finallg arriving in Karachi to begin our assignment.*
>
> *Later we went to Lahore and the Falleti Hotel. Then to East Pakistan's Dacca University where I spent the semester lecturing to a class of graduate students. My lectures were on colchicine. During the process of preparing the lectures, I discovered that our manuscnpt was not well written. If you want to test a good manuscript, just read it to a class. Being totally dissatisfied with the writing with one mighty heave I threw the entire manuscript into the fireplace and the next morning started writing a new one from scratch.*

There were many activities that allowed O. J. and Agnes to rub elbows with a society that enriched their notions of how humanity lives. The houseboy's assurance that Allah would not let the Eigstis become ill whether or not he washed the dishes as Agnes was accustomed to doing is only one of many such incidents. A more professional extension of the initial Fulbright concept took O. J. to India's University of Lucknow to serve as a UNESCO lecturer at the 40th Indian Science Conference. This was done durning vacation time and led to a life membership in Phytomorphology, Botanical Society of India. In fact, this and many similar incidents bear witness to the positive features that the Fulbright program added to the O. J. story.

En route home from the Fulbright assignment, O. J. served as Research Consultant on colchicine at the University of Brussels, Belgium, and at the University of Lund, Sweden. By now, the Eigsti sons had rejoined their parents and there were many richly rewarding experiences for everyone while O. J. continued his professional activities. But there was nothing upbeat among the circumstances the Eigstis faced on returning to the United States. Changes in the administrative setup at the Funk Brothers Seed Company left O. J. virtually unemployed despite the Fulbnght policy requiring his full renstatement. The approaching holiday season of 1953, rather than being a festive homecoming, is remembered as the lowest, most painful of all of the periods of deep depression that have been a part of O. J.'s half century mission to make the seedless walermelon a financial success.

These and other circumstances discussed elsewhere prompted O. J. to return to teaching. The first acceptable job opportunity that came to his attention was at the Chicago Teachers College. The position included teaching botany and genetics. O. J. continues:

> *It was exactly what I was trained to teach. The salary exceeded that of my colleagues at other universities and there was extra pay for time beyond fifteen contact hours. Unheard of in my other teaching jobs. And those other jobs became a bit embarrassing. During the interview, I sensed there were questions as to why I would be interested in a position that seemed to be a U-turn on the way up the status scale. This bothered me. I would be teaching would-be teachers. I let my potential employer know that I felt no let down professionally.*
>
> *I was hired on a trial basis. If I proved to be satisfactory, I would be recommended to take an exam that all teachers in Chicago's schools had to take. It wasn't an exam on the knowledge I was to teach but it was a bunch of silly nonsense. Well, I passed the exam. I must say educatian courses et Bradley and Goshen did indeed help.*

So it was that in 1955, O. J. began what became a twenty-one year stay at the Chicago Teachers College. When the name was changed to Chicago State University three years later, he received the rank of associate professor. He advanced to full professor in 1959, and retired as emeritus in 1976. His position gave him a wonderful opportunity to observe and to take part in highly significant changes in genetics and how all branches of science are to be taught.

My explanation of O. J.'s success as a teacher is admittedly simplistic. O. J. taught as he was taught to teach. This brings us to focus on two parameters. The first of these is the methods he used, the educational psychology undergirding his presentations, and the values reflected in relationships with students. The second is his control of and attitudes toward the subject matter he taught. The period between 1955 and 1976 creates a unique opportunity to examine how one teacher responded to unparalleled ferment in both of these parameters.

Beginning with genetics-related course content, when James D. Watson and Francis Crick unfurled their double helix model for DNA in 1955, geneticists everywhere found themselves with more than just a new vocabulary to learn. The primary focus was no longer on chromosomes and genes as discrete objects. In keeping with the spirit of science to seek increasingly detailed descriptions of nature's events and circumstances, it was no longer enough to observe what genes are and can do. Based on the fundamental principle that the kind, number, and arrangement of its component atoms determines what a molecule can do, the prevailing challenge became to trace the action of each gene to its unique atomic structure.

It is to be expected that this change in focus would affect O. J. If he was to bring his students up to where genetics was actually being pursued, he would have to assume the role of student simultaneously with that of teacher. In practice, this dual role worked to everyone's advantage. A primary reason was that O. J. was familiar with just about all of genetics-related knowledge that had accumulated prior to the advent of DNA. Equally fundamental, he knew where genetics fits in humanity's urge to understand the world around and within us. No matter how elegantly the DNA substance was portrayed in the popular media or how elaborate the audiovisual aids that were developed to convey its amazingly intricate atomic structure, always in the back of O. J.'s mind was the question of what genetics-related events might be better understood, managed, and utilized by calling on the new knowledge made available by Watson and Crick.

Success as a teacher, however, depends upon things that reach beyond what is in teachers' minds and what they want to have happen m the minds of their students. There is a world of finesse in recognizing each of the experiences that become synthesized into the ultimate appreciation of a bit of nature. Developing, nuturing, and honing the skills required to assemble separate glimpses into the whole of a naturally occuring event requires great pedagogical skill. All of this can be done best by men and women who have matched their own minds and hands against whatever are the forces that seem to protect the ways of nature from humanity's prying eyes and meddlesome hands.

Without doubt, O. J.'s students heard much more about seedless watermelons than they expected to hear. But it was for a very sound pedagogical reason. And not even O. J. can say for sure that the ultimate grasp of the right solutions to at least some of the problems he faced in fields of watermelons originated in the intellectual give and take in those totally urban classrooms occupied by naive geneticists.

Much has been written about science teachers at the collegiate or university level being handicapped because many of their students come to them lacking basic experiences that enable them to grasp or truly appreciate accurate descriptions of the event under consideration. Early on, O. J. decided to not dwell on, but to compensate for such an admittedly frustrating handicap. In a sense, he treated the problems to be dealt with in the classroom pretty much as he did those he encountered in the pursuit of science.

Much has also been written about the problems that students bring with them that stem from their ethnic backgrounds and from home and community socioeconomic conditions. When O. J. came to the Chicago Teachers College, he brought a wealth of experience working with people of widely diverse ethnic origin and economic status. Growing up in the Mennonite faith, he was not a stranger to being discriminated against by people whose religious beliefs permitted such discrimination. The seven years at the University of Oklahoma kept him in touch with the moral and social values that characterize people who are proud of a way of life that integrates America's Southwest with the far West, the deep south, and the one-time prairie country.

The move from Oklahoma to Northwestern University increased the dimensions of O. J.'s sensitivity to diverse ethnic origins and socioeconomic attributes. These were times when empathy was needed throughout the science teaching profession. It was deeply involved in reform movements. The so-called post-Sputnik era was marked by widely divergent notions of the what, how, and why of the science curriculum. Large amounts of federal money were being invested in rewriting textbooks, redesigning laboratory exercises, developing instructional aids, and orienting novice and experienced teachers in new ways to treat familiar subject matter and handle added new material. As the name, Chicago Teachers College implies, O. J. recognized the need to not only give his students a thorough subject matter background, but to also practice efficient teaching methods.

How O. J. acquired his ways to teach is difficult to establish. I begin with evidence provided by his extensive underlining and marginal notes in the textbook, *Principles of Secondary Education* by L. A. Williamson and G. A. Rice and published by Ginn and Company in 1927. Because they highlight information that continues to be valid, it may well be that O. J.'s competency in the classroom began with learning the fourteen "laws of learning" that were "cautiously advanced" by this textbook's authors.

1. Learning takes place only in activity. This learning may be mental, emotional, or physical, positive or negative, but to be actual the learner has to participate.
2. Both immediate and ultimate purposes must be known to the learner if learning is to be effective.
3. Speed is essential to good learning, but the learner should set the pace, or at least the rate should conform to his capacity.
4. There are certain physiological limits to improvement which should be studied for each individual and each group. The first of these is inherited capacity, and it should be recognized that this is no fault of the pupil.
5. Practice with satisfaction is the shortest road to permanent learning. Success is the most important factor in stamping in a learned response, for there is no human law that is more universally recognized than that man tends to do what has been successful in the past.

6. Coercion, founded upon the psychological factor of fear, is a poor motive force for education. Especially does coercion develop a bad mind-set, a mental sabotage, a tendency toward deception, an attitude of condoning 'anything that one can get away with,' which is having a bad effect upon community life; nor is it efficient, compared to purposeful activity, as an educational motif.
7. Interest is the psychological factor upon which learning should be founded. Enlightened self-interest has the correct emotional drive. There is more discipline, more practice in overcoming obstacles, more concentration on the work in hand, more shutting out of thwarting influences, more practice in persistence, originality, initiative, and all of the other virtues in the calendar, in following one's own bent than there is in going against one's own interests.
8. Subject matter is primarily a means and only secondarily an end.
9. Subject matter should be presented as a problem.
10. The act of learning is psychological rather than logical. The personality of the pupil, his attitudes, his purposes and interests, have more influence than logical modes of thought.
11. The act should be learned the way it is going to be used. Whenever there is a technique, a skill, or a habit to be learned, it is essential to teach it exactly as it is to function.
12. Learning is more than an intellectual process. Learning that does not result in change in character or added skill, or some useful thing or act, is not true learning—it is mere scholastic activity.
13. Attitudes and ideals function only when practiced under all conditions. One learns to be dull in a dull classroom, one learns irresponsibility when no responsibilities are thrust upon one, one learns apathy where interest is lacking, one learns mediocrity where one's talents are not exercised.
14. Understanding and good judgment come from wide use under a great variety of conditions.

Concurrently with O. J. returning to teaching, biologists and biology teachers were struggling with where the emphasis should be in teaching the life sciences. Financed by the National Science Foundation, three new high school biology textbooks became available. The Green Version continued to feature life as it functions in the natural environment. The Yellow Version's principal focus was on the genetic evolution of life's myriad forms. The Blue Version, often identified as the molecular version, was heavily influenced by the challenge to reduce all of life's attributes to changes in the properties of substances that result from the reshufflings of atoms from molecule to molecule.

In effect, the first version was to satisfy those who believed in introducing young people to the efficiency of life's processes, the intricacies in the life histories of representative species, including, of course, our own, the interdependence of species; and the impact of environmental factors on the past, present, and future of the countless forms in which life appears. Teachers who see their primary mission in the classroom to be to develop and enhance appreciation of the origin of all species would choose the second version. The third version would be preferred by those who are particularly challenged by searching out all of the chemistry that makes possible what it means to be alive.

It is noteworthy that O. J.'s work on seedless watermelons put into practice all three of these interpretations of what biology's identifying features should be. It is to be expected, therefore, that he would integrate all of them at the lecture podium and in the instructional laboratory. And to be impatient with those who proclaimed the superiority of any one version at the expense of the others.

While Eigsti was growing up in science, life was generally accepted as being an entity, a property of protoplism. By 1955, the year O. J. returned to teaching genetics, highly pervasive fotces within the scientific community began interpreting life as process, a smgle or integrated set of complex chemical reactions. No other divergence of focus could have such enormous impact on biology course content. Much of this change grew from the work of Harold Urey and Stanley Miller. They hypothesized that life is a chemical process that resulted from the coincidence in time and space of the requisite reactants and reactant conditions. Urey and Miller then devised an experiment that brought together their best estimate of what these reactants and reactant conditions must have been. When the experiment was concluded, the reactant vessel contained products that are essential to and can be seen as precursors of the ultimate life process.

Publication of the results from this experiment quickly permeated the thinking of the world's biologists. Less rapidly, but none the less completely, the implications that could be drawn began influencing what biologists were doing in their laboratories and what they believed should be taught in school and university biology classes. It is to be expected that this would lead to controversy that would surely affect what O. J. would be expected to teach and the way it should be taught.

Early in O. J.'s educational experiences, the ways and means of science, the methods scientists use, were considered an essential part of the science curriculum. Science courses included laboratory exercises that were designed to train students in gathering data, formulating hypotheses, and designing, carrying out, and interpreting the results from experiments. At the secondary level, the daily program provided "double periods" for work in well-equipped laboratories. At the time when O. J. reentered teaching, other interests were concerned about emphasizing the teaching of science as process and questioned providing adequate facilities to give students practice in approaching problems scientifically.

This questioning held no threat to O. J. Eigsti. The strengths and limitations of the ways and means of science were a part of his life. They needed no justiflcation. On the contrary, they became motivation for his students. O. J.'s ability to flesh out science-related knowledge reminds me of lessons taught by James B. Conant, a student of both science and education while he was president of Harvard University. To Conant, the teacher in the classroom and the student in the laboratory both boil down to one intellect, often the "glorious amateur," wrestling with the demands of the moment.

There is another element in O. J.'s education that is equally difficult to convey. He moved easily among people of all stations and persuasions and cultivated meaningful conversations. He was quick to defend those whose work and points of view he respected. This is well documented by marginal notes in the previously cited *The Situation in the Biologocal Sciences: Proceedings of the Lenin Academy of Agricultural Sciences of the U.S.S.R.* O. J. took personally Lysenko's fabricated, slanderous, and perverted accusations leveled against geneticists whose lives and work he respected.

But the climate O. J. fostered in his classroom was not that of the ivory tower. In his own way, he remained sensitive to his student's needs and interests. This is difficult to illustrate, but one often repeated circumstance comes to mind, and serves well to end this phase of the life of Professor O. J. Eigsti.

> *At Chicago Teachers College when I started there in 1955, I always brought seedless watermelons to my first lecture. They made a nice introduction to genetics. Then we would cut the melons and I had in my classes black students and white students. The white students would kinda turn up their noses, but I never found a black student who didn't like watermelon and after it was all over, they'd stay and clean up the mess. Every one of those who stayed was black. This business of black people not wanting to be identified with watermelons, I never saw it. As time went on, I got more black students than white and finally, twenty years later, eighty-five percent of my student were black.*

Chapter 9

Solving Financial Problems

THE STORY OF the Eigstis solving financial problems begins with the gold coins sewed in the skirt of an ancestral aunt as she came to Ellis Island in 1850. These coins financed the purchase of an Illinois farm, and they are symbolic of much more being brought from the old country to the New World. With them came the tradition of producing things, growing things that one's neighbors need and are willing to pay for, and being free to accumulate the profit. These values became a part of O. J.'s life as he grew up while joining other family members in peddling their farm produce on the streets of Peoria. At the same time, his father, vice president of the neighborhood bank, taught O. J. valuable lessons from the financial world.

The second milestone, coinciding as it did with O. J.'s eighty-eighth birthday, marks what could well be the most keenly felt of all of the stings and bruises he has experienced in his financial deals—a $90,000 claim from the Internal Revenue Service based on the previous year's tax liability.

What it is appropriate to say here may not truly convey the Eigstis' attitudes toward money—attitudes rarely verbalized, but often expressed by their actions. To them, money has value only in what it enables one to do. In itself, it is worthless. Its possession is meaningless. Nothing makes less sense than to play out one's life struggling to get more of it only to stash it away. Nothing is gained by flaunting symbols of possession of immense quantities. To the Eigstis, the lack of ready money is not permitted to deny them the opportunity to explore a promising idea or hypothesis or to obtain essential wherewithal. For example, only the yield of valuable information, not travel cost, is to be taken into account in deciding whether or not to travel to Japan, Paraguay, Belgium, or anywhere else.

Recognition of their attitudes toward money is essential to understand and appreciate O. J. and Agnes Eigsti, and what they stand for. If they lived in the biggest house on the street, it was because there was garden space to grow albino soybeans, raise strawberries, and conduct cross-pollination experiments. If they spent thousands of dollars to buy an adjacent lot, it was not to reap profits from a real estate investment, but to enable Agnes to continue to enjoy the environs of her family home. If O. J. served a very expensive brand of Scotch whiskey, the contents meant more than the label on the bottle. It is to be expected that such attitudes would sometimes confuse friends and neighbors. It would be difficult to sympathize when setbacks occurred while the Eigstis apparently continued to afford a more than comfortable lifestyle.

Money-related anecdotes abound in the Eigsti ancestral history. There is the story, mentioned above, of one ancestor arriving at Ellis Island in 1850. It was there that the name Augstien was changed to the more easily spelled Eigsti.

> *We learned of this Augstien name in a very interesting way. There were three brothers and a sister living in Bavaria. Their sons were becoming eligible for military service so two of the brothers and the sister came to America. One brother went to Canada, the other brother and sister came to Morton in Illinois. Actually, the family originated in Strasburg and later moved to Bavaria because after the Thirty Years War, Germany was devastated and Mennonites were known to be good farmers. And so Nickolas and his wife were invited by the Duke of Bavaria to restore and rehabilitate the countryside. In the process they must have accumulated land so when the sister arrived at Ellis Island she had $3,000 in gold, some say it was $5,000, sewed in her skirts. That much gold was quite an accumulation in the 1850s and it enabled her to buy the land near Morton which was farmed by my grandfather, Christian Eigsti, and his brother. We don't know how they escaped the Civil War, but in those days you could buy your way out by paying someone else so much to go in your place. That was customary. They wouldn't have gone willingly. They came to the United States because of the militarism of Bismarck. They were pacifists, Antibaptists, Mennonites.*

One might assume that although financial security prevailed among O. J.'s ancestors, the Great Depression would burn its sense of insecurity deep into the persona of anyone coming of age during the '30s.

> *We suffered during the Depression there on the farm. But we were better off than a lot of other people. We at least had food. But then,*

I remember that after I had stayed out of school for a year, my father and I sat down on a swing and he said to me, "I think you ought to go to the university. Would you like to go to the university and become a teacher?" Sure, I said. Mahlon, my brother, had just graduated from Bradley University. My father wanted me to go to college. It was his idea, not mine. He figured I shouldn't waste my time being a farmer. I was a pretty good student. I wasn't a dummy. A teacher had given us the Otis Intelligence Test and he told me what my IQ is.

There are other reasons to believe that O. J. escaped the financial trauma that handicapped so many of his generation forever after the Great Depression. The Eigsti family bought their first automobile in 1911, the first in their community, but this didn't keep them from using it to haul the farm produce they peddled on the streets of Peoria each weekend. It is also significant that O. J.'s father was a vice president of the Morton bank for nineteen years.

My father held his money very tightly. He never ran into debt. No, we didn't have it so bad during the Depression years. We never borrowed money, even though he could have borrowed all he wanted to. He could have bought thousands of acres of farm land for a hundred dollars an acre that is now selling for thousands. He was very frugal. But he was a spendthrift compared to his father. My grandfather, Christian Eigsti, came to Morton, Illinois, penniless. And when he died in 1903, he owned one thousand acres, paid for.

When I got to New York and saw those bread lines, those soup lines, in 1933, '35, '36, and '37, I realized we were pretty well off. Nor did the dust storms ever hit us. We never had a crop failure. We had a good farm. That land there in Morton was good land, the best land in the world.

Money problems loomed large throughout much of O. J. Eigsti's life. Solving these problems consumed a large share of his time, energy, and spirit. But in their solutions lies a full measure of what it has meant to be a scientist, a teacher, and an agriculturist throughout the twentieth century. This is a source of insight and inspiration for those who choose to factor the past into designing solutions to financial problems as they evolve in the future.

The Eigsti life story retraces step by step how research people solved financial problems during the twentieth century. Again, however, such problems rarely invite open discussion, no matter how heavily they bear on a person's decisions. It is best to treat this facet of O. J.'s life "generically" and let it be seen as providing guideposts along the way toward career decisions.

At the beginning of their careers, researchers often recall particularly empathetic university or collegiate instructors judiciously allowing them to consume more than a fair share of the department's resources. In fact, there have been times when an "unauthorized" reagent or specialized apparatus was purchased from department funds to accommodate a deserving student. When students were paid to do essential departmental chores, they received added dividends ranging from developing valuable friendships to learning to maintain trouble-prone apparatuses.

Campus jobs developed interpersonal skills in addition to keeping body and soul together. In O. J.'s case, a job as director of physical education while he was a student at Goshen College led to lifelong friendships. The position assistant in botany at the University of Illinois helped cover the expenses of his master's degree and led to a summer job as assistant cytologist at the Carnegie Institute of Washington's laboratory at Cold Spring Harbor. Advancement to the rank of fellow in botany at the university solved some of the financial problems that came with earning his Ph.D. degree.

Associate in genetics at the Cold Spring Harbor laboratory, O. J.'s first full-time job as a scientist, was representative of the few available entry-level positions during the economically depressed conditions of the late 1930s. Viewed in retrospect, the rare good fortune represented by his receiving this appointment bears witness to his having made a good impression during his summer job at the laboratory in 1933.

The part-time position with Triarch identifies several of O. J.'s job interests. He was free to pursue research of his own choosing. He believed in the value of his employer's product. He was using and improving a mastery of skills that he held to be worthwhile. At a more subjective level, he was gaining first-hand experience with the ups and downs of entrepreneurship as the market for instructional materials in biology waxed and waned.

In 1937 Eigsti considered it to be very good fortune to receive an interim appointment as professor of biology and director of physical education at Greenville College in Illinois. This was ten years after starting college with the goal of becoming a teacher. Knowing that this position would not be available beyond the end of the year, he applied for jobs elsewhere and received several acceptances. He chose the position as assistant professor of botany at the University of Oklahoma. He advanced to the rank of associate professor in 1940 and to professor in 1944.

With this appointment, O. J. took on a way to solve financial problems that has become common practice among university professors. His position with the university allowed him to pursue topics of his own choosing and the "professor" identity invited offers to serve as "consultant" to agencies in a position to benefit from "expert" advice. It was to be expected that O. J.'s training and experience in genetics would invite such offers from the rapidly emerging hybrid corn industry. For example, he developed a variety of corn that was uniquely adapted to the Oklahoma growing season. His work with farmers in their fields kept alive his feeling of being at home with farm people.

After seven years at the University of Oklahoma, O. J. accepted a position as associate professor of botany at Northwestern University in Evanston. It is seldom easy to look back and articulate the reasons why people leave one job and take another position in the academic world. Without identifying the drummer's beat that O. J. chose to listen to, in 1950 he accepted a position as research botanist with the Funk Brothers Seed Company in Bloomington, Illinois. This career change has special significance. Bloomington is only across the street from Normal, Illinois, a geographical state of affairs that led to the Eigsti and the Woodburn families becoming not only neighbors for the next three years, but friends ever after.

When O. J. joined Funk Brothers, the hybrid corn industry was well past its adolescence. The science supporting the industry had been pretty well worked out, as had the technology. But there were still potentially devastating problems. One such problem grew out of hybrid seed producers being dependent on hiring people to laboriously remove the tassels from the rows of corn that were not to supply the pollen for the ears developing on the intervening rows. The discovery of the male sterile trait in an experimental seed line promised to eliminate the costly detassling process. After being rather widely promoted, however, "side effects of the male sterile gene made this solution to the detassling problem a total failure. On the other hand, the genetics of the parent seed lines had been worked out to the point where hybrid vigor ensured that insects and disease-resistant varieties would produce high yields.

This is not to say that there were no longer challenging opportunities for geneticists in the hybrid corn industry. At one point, dozens of varieties of corn were maintained as a "germ plasma bank" in the Funk test plots. At that time it had not been brought to the public's attention that each relative of the corn plant, no matter how unpretentious or unproductive it appeared to be in its native habitat, might have in its genome a trait that would add disease resistance or an equally valuable trait to a "domesticated" variety of corn. Brought to Illinois from all over the world, the genes in these "wild" relatives promised to contribute to the success of America's corn industry.

Almost from the beginning of his role as an "industrial" geneticist, O. J. was troubled by a feeling of not being entirely free to pursue his own interests, to practice his own notion of social values, to go wherever and to do whatever he believed would make his life count for something. Needless to say, the potentialities of polyploidy as applied in the production of seedless watermelons were deeply embedded in his hopes and ambitions. And there was the equally persistent urge to publish the complete story of colchicine's role not only in causing polyploidy but in all of nature's phenomena.

After two years with Funk Brothers, O. J. asked for a leave of absence and applied for a position as Fulbright professor and visiting professor of genetics at the University of Dacca in East Pakistan and the University of Punjab in West Pakistan. This government-sponsored Fulbright Program became the solution to financial problems for 1953–54.

Nothing happened while O. J. was in Pakistan to relieve discontent with his position at Funks, and it was mutually agreed that he would leave. The Eigstis were now free to organize the American Seedless Watermelon Seed Corporation and the Colchicine Research Foundation. This marked a highly significant change in finances. No longer was there income from other sources to solve family budgetary problems and finance further efforts to make seedless watermelons a commercial success. Central to this new solution to financial problems was the issuance of $15,000 worth of stock in the newly organized corporation. Sales of seedless watermelon seed were projected to yield significant income—and there were sidelines. Both Agnes and O. J. continued to produce instructional microscope slides for Triarch. Also, they raised and sold, both retail and wholesale, a variety of soybeans that carried the albino trait and "Indian" corn that carried special genetic traits.

Cash flow, however, soon became a serious problem. Income simply would not support expenses. In 1955 O. J. accepted a position at the Chicago Teachers College. Agnes was employed as head of the parasitology department at the General Biological Supply House making instructional microscope slides that were sold under the Turtox label. The Eigstis maintained a residence in Chicago, but spent as much time as possible at their farm home in Goshen, Indiana.

The decision to issue and sell stock in the American Seedless Watermelon Seed Corporation demanded reconsideration. Friends held most of the stock. There were no profits to be distributed as dividends, a state of affairs that caused the Eigstis to feel a sense of disquieting anxiety. A fortuitous family inheritance enabled them to buy up all shares of the stock. In addition to relieving a sense of debt to friends and relatives, this also enabled them to gain total control of the corporation.

In 1976 O. J. retired from Chicago State University and Agnes gave up her position at General Biological. The family income was reduced to retirement pay and sales from the Seedless Watermelon Seed Corporation. Again, although the seed production business was expanding, cash flow problems demanded solutions, At one point, significant shares in the corporation were sold to relatives. By 1981, the financial condition of the corporation had deteriorated to a condition reminiscent of that which prompted repurchase of all outstanding shares of the stock. Again, a sense of fair play prompted the Eigstis to borrow $80,000 and regain full ownership of the corporation.

Before continuing with O. J.'s experiences in the financial world, it is noteworthy that the Eigstis' solutions to financial problems include sources of funds that, in themselves, help us to understand O. J. and also illustrate how the scientific enterprise operated during the twentieth century, particularly the first half. Identified as "grants in aid," "research awards," or "travel grants," O. J. received financial support from the University of Illinois, the Works Project Administration, the University of Oklahoma, Sigma Xi, Oklahoma Academy of Science, the Oklahoma Research Institute, the Carnegie Corporation of New York, Northwestern University, the Genetics Society of America, the American Cancer Society, National Institutes of Health, the University of Brussels, UNESCO, the Swedish Institute of Genetics, the Iowa State College Press, the Polyploidy Crop Association of Japan, and the Chicago State University.

The wide geographical range of these awards and grants tells us that O. J. Eigsti's work has been recognized worldwide. The status of the granting agencies tells us that he is a "scientists' scientist," that he can speak their language, share their motivations, and play their game. O. J.'s experiences in his first full-time position as a scientist, which were sufficiently traumatic to be remembered with emotional emphasis ever after, make it clear that all of this could easily have been otherwise if he'd left the field. For a scientist with a new Ph.D. to separate himself from one of America's most prestigious research institutions in the midst of the greatest of all financial depressions could easily have spawned fatal interpersonal shortcomings. That this did not happen bears witness to the integrity of not only the individuals who were involved, but to the positive features of the scientific community.

It is also to be noted that the Eigstis' solutions to financial problems lacked one feature that became dominant during the second half of the twentieth century. Establishment of the National Science Foundation and its increasing financial support by Congress offered a widely popular solution to researchers' money problems. Why the Eigstis were never able to take advantage of NSF grants during the research and development phases of the commercialization of seedless watermelon remains hidden in the sometimes obscure circumstances that determine what applicants are willing to do to obtain grants and what reviewers require on the part of the grantee.

Actually, trying to retrace the paper trail of the Eigstis' solutions to financial problems that had to be solved before the seedless watermelon became a financial success tells only a little about hundreds of paychecks earned. It doesn't call to mind the instructional microscope slides, albino corn and soybean seed, and Indian corn that had to be produced, advertised, and sold. Or what had to happen before stock certificates could be issued and sold to friends and relatives, only to be bought back when no dividends were forthcoming. In reality, money problems do, indeed, loom large in O. J.'s memories.

A period beginning in 1976 was exceedingly gloomy. Both O. J. and Agnes had retired from their positions in Chicago. O. J. was now free to give all of his time to the seedless watermelon enterprise. Seed production had expanded, particularly in Texas. Unfortunately, expansion of production and sales increased the dimensions of losses when bad things happened. One expensive example was the sale of 250 pounds of seed to a Florida grower. After the cost of planting all of these seeds, not a single one germinated. The cause was traced to procedures used to separate the seeds from melon flesh and rinds.

Ordinary watermelon seeds are separated by allowing the crushed melon flesh to ferment. It took some time for O. J. to discover that because triploid seeds lack certain internal structures, water would seep into the seeds and "drown" the embryos. Separation of triploid seed required hitting the crushed unfermented melons with a high pressure water jet. This is only one of the expensive setbacks suffered during the late 1970s and early '80s. In fact, it was during these years that a sense of responsibility caused O. J. to borrow $80,000 to buy back the stock owned by relatives.

Not only was this contrary to financial advice, it required scraping up all available collateral. But O. J. remembered how Joseph and Robert Weaver had gone to simialar ends to keep the business afloat. In fact, Joseph Weaver managed the business for twenty-one years in a most conservative manner, another way to say he kept the business from going bankrupt. And perhaps more indicative of family morals than of money matters, nieces and nephews added their support by hand-pollinating the 4008 tetraploid which continues to make a very important contribution to the watermelon industry.

The final phases of O. J.'s solutions to financial problems brought together such a wide array of seemingly random circumstances that they appear to be more fanciful than real. Some reach back to his travels during sabbatical leaves. For others, there was no hint that they would happen until well into the 1980s. But there was nothing fanciful about the Eigsti finances in the mid 1980s. Setbacks had reduced O. J. to the lowest possible level of confidence in the ultimate commercial success of the seedless watermelon.

> *I tried to sell the company for $100,000. Nobody wanted it. Everyone said there was no future for seedless watermelons. We reached an absolute low. It was the Fourth of July, 1985. Karl came out here on a visit. He asked, "How is the seedless watermelon business?" The seedless watermelon is nothing. I am going to take all of my tetraploid stock and put it in a pile, pour kerosene on it and burn it. That's one*

way to get out of this business. It isn't anything. Karl said, "You can't do that." That's what I am going to do. I am absolutely at the end of the line and I never want any more to do with seedless watermelons.

Obviously, this wasn't the end of the story of O. J. and the seedless watermelon. But it marked the beginning of a feature that is characteristic of the evolution of science-related events in agriculture during the second half of the twentieth century. No longer was it practical for farmers to rely on their own financial resources to develop new ideas. In O. J.'s case, he faced two options: either sell the business or look for additional financing.

At this point, his son Karl decided to help O. J. keep the American Seedless Watermelon Seed Corporation alive. In his work as a set designer for stage productions on and off Broadway, Karl was quite familiar with the ways and means of financiers. He arranged with Joe Rubin, a professional entrepreneur, to help solve O. J.'s financial problems. As remembered by O. J.:

So in February of 1986 [Joe] called me to come to New York. We met in his office. He wanted to organize a group to continue what I was ready to abandon. I said we needed three growers to produce the seed and I wanted the family in it. He drew up what is called the shareholders' agreement. It's got all kinds of stuff in it about what we can do and can't do. Joe Rubin is really a very good lawyer. He's sharp! Karl then became President and I was Chairman of the Board. Joe Rubin is Secretary. Nick had other things to do, so Agnes received his share. So we got Bruce Price, Bob Shipley, and Donald Johnson for the three growers.

Among the many watermelon growers in the American southwest, it is interesting how O. J. "recruited" these three growers. It was again pretty much a matter of random circumstances. Bob Shipley was well established in the wholesale watermelon business in Nogales, Arizona. In 1982, he came to Texas, where growers were producing Eigsti's seedless melons. He wanted both seedless melons and the Jubilee variety of regular melons being grown by Donald Johnson.

When Bob went to get the Jubilee melons, he asked Donald why he was not growing seedless. Donald said he would like to but was unable to get seed. It seemed that the growers wanted to keep their seed for themselves. This is not unusual in the watermelon business. Then Shipley told me about Johnson, so I went to his farm in the spring of 1983. We planned a small production to get him started.

And then came bad luck. In the summer of 1980 and again in 1981, I was getting tired of poor help, poor equipment. We had our good seed processing machine but it was back at Havana, Illinois, so we borrowed a seed harvesting machine. This turned out to be a disaster. It had been used to process regular seeded melon seed and we didn't clean it. The washer was loaded with Chubby Gray seed and the seed we sold the Johnsons was contaminated. To make matters worse, there is no way to tell Chubby Gray melons from our seedless in the field. Fortunately, we discovered the problem and could rely for the next year on seed produced at Havana.

In 1984 we used our own equipment. The Johnsons had a good crop of seedless melons. Not only were the Johnsons good to work with, facilities at their ranch were so much better for washing, drying, and so on that we came to rely on them more and more. We became well acquainted and I spent many pleasant hours around their table. When I told them I was ready to throw in the towel in 1985, they were among the few people who kept my spirit alive.

From Donald Johnson's point of view,

For the last thirteen years, the Johnson family has had a very close and special relationship with the Eigsti family. The Eigstis have spent much time with us in Knox City, Texas. They have had a tremendous impact on the Johnsons' lives. O. J. not only helped with harvesting the seed, he also taught the necessary processes to the company employees. Due to high-risk weather problems, the seed production has been moved to California, but each year we now grow 600–800 acres of seedless watermelons in Knox City. The variety, Tri-X313 is such a superior product, no one has been able to improve it. It is the #1 seedless variety in the world. Today, [we] have much love and respect for O. J. and Agnes Eigsti. We feel so lucky and fortunate to have become involved with such wonderful people.

Well into the 1980s, little happened to bolster O. J.'s confidence in the ultimate commercial success of the seedless watermelon. At one point, he and his son Nick, in what could have been a last-ditch effort, set out to survey the seedless watermelon situation in California.

That's right. We visited every one of the important growers. We ran into one man who was very enthusiastic. "They jump off the rack." The man who said that was Harry Otto, who was an extension agent for Orange County. He had nothing to do with the seed business. They were producing seedless melons at the Irvine ranch. Here is a part of our story that has never been told and it is very important. Harry Otto and Roy Kobayashi were in the Extension Service for Orange County. They were doing a lot of work with the Irvine ranch.

The Irvine ranch is a tremendous holding of which Howard Margoulis has a big interest. He's president of Sun World, the enormous produce wholesaler. He doesn't own the ranch but he has a big interest in it. Harry Otto . . . started growing seedless melons as a high school student. After he graduated from Michigan State, he went to Imperial County, California, to work in the Extension Service. When he got there he said, here is the place to grow seedless melons. So he got in touch with us and wanted seed. We probably gave him the seed. OK, they don't germinate. But Roy Kobayashi . . . learns how to get these seedless seeds to germinate. He raises seedlings in the greenhouse and transplants them to the field.

Later Harry and Roy moved to Orange County, the jumping off place in the world for growing watermelons. This is where he got involved with the Irvine ranch. For years they kept growing seedless melons. This went on for ten years. Then one day he phoned me and said I got a call from Howard Margoulis at Sun World and they want to get into seedless melons in a big way, they want to grow 200 acres. They're all excited. But Harry said he didn't think they should take on that much. We better limit them to forty acres. I said absolutely. Harry Otto was the key here. Instead of going for 200 acres, we limited them to forty and they were tremendously successful. And that success really stimulated Margoulis! He was really excited.

So, in 1987, Margoulis and his associates from Sun World came to a meeting of the National Association of Watermelon Growers at San Antonio. There were research workers from all over the world. After this meeting, Joe Rubin had Margoulis and some of his people meet with us in New York where we whipped up an agreement that Sun World and American Seedless would form a partnership for growing and selling seedless watermelons. The reason it is called the American Sunmelon partnership is that the American comes from one corporation and Sunmelon from the other. And Sun World has outlets all around the country where they sell peppers, dates, and all kinds of fruits and vegetables. Actually, their outlets are worldwide. And I'm telling you, our business just went to town. That was a success. And if it hadn't been for Sun World, we would probably have no more than struggled along.

There was much behind the scenes "where we whipped up an agreement" that O. J. doesn't mention. When Karl returned to New York after that July 4th visit, O. J.'s threat to abandon the seedless watermelon weighed heavily on his mind. Many months before O. J. went to New York, Karl tried to sell the American Seedless Watermelon Seed Corporation to existing seed companies. None was willing to pay the asking price. But Karl sensed increasing interest in the seedless watermelon—interest that told him to not lower the price. One circumstance was particularly impressive. Growers in Israel learned that by using special irrigation and mulching methods, they could produce up to 30 tons of seedless melons per acre. When these melons were sold at premium prices, there was no doubt about their commercial success.

At the same time that Karl was gaining confidence in the future of seedless watermelons, he also realized that he needed help to deal with financial problems. This led to the arrangement with Joe Rubin, a lawyer with wide entrepreneur experience. The arrangement with Rubin proved to be very effective. He not only advised Karl during his efforts to sell O. J.'s company, but worked up a fall-back plan if no sale was forthcoming. This plan arranged for the development of the previously mentioned American Seedless Corporation. Although this new corporation continued to produce, process, and sell seedless watermelon seed, cash flow problems kept alive the plan to seek a better solution to financial problems.

A major step in the direction of such a plan began while Karl was in San Diego pursuing a sales prospect. A phone call from Carl Sam Maggio, co-owner of the Sun World produce company, led to a business dinner meeting. In typical fashion, although Karl sensed Maggio was interested in the seedless watermelon enterprise, no details were discussed. But the previously mentioned meeting with Joe Rubin in his New York office was scheduled.

This meeting, in turn, led to a second meeting at which both Maggio and David Margoulis represented Sun World's interests. Development of the American Sun Melon Partnership was the outcome of this meeting. In effect, Sun World put up $600,000 for 50 percent of the American Seedless Corporation, which enabled that corporation to pay O. J. $250,000 for his share.

When we put these East Coast and West Coast events together, they become the culminating action of the financial part of the drama that is the Eigsti family's story. Within these closing scenes, we see things happening that illustrate fundamental changes in rural America during the twentieth century. Solving financial problems for the Eigstis began as a family enterprise, self-contained, self-managed, and self-financed. But this was not possible as the century came to a close. When Sun World paid $600,000 for all domestic rights to the American Seedless tetraploid seed stock, the Eigstis reached the threshold of the twenty-first century.

Changes accompanied the transition—changes that are characteristic of many phases of the rural American society. Their share of that $600,000 enabled the Eigstis to pay off all of their debts and freed them from financial worries. But their interest in the seedless melon was no longer self-managed. Sales of seedless melon seed, although permitted overseas, were now limited to Sun World's growers. This caused concern among other growers, but other seed companies, benefiting from the boost that came with Sun World's stimulation of sales of seedless melons, reconsidered their attitudes toward seedless watermelons and were soon able to more than meet the demands for seedless melon seed. For the

Eigstis,

> *That was the beginning of our financial improvement. And we had a good relation with Sun World. Why my goodness, this last year, every year they have earned dividends equal to what they put in that first year. I'm telling you I kept my fingers crossed. I didn't know whether it was going to go or not. I had so much failure over the years, those '50s, '60s, and '70s, all I knew was failure.*

Once more the O. J. Eigsti life story is best seen as a small window through which to view big things happening throughout the twentieth century. At the beginning, we have the Larsons selling their seedless melons at a roadside stand. At the close of the century, there is Sun World where many salespeople make the necessary connections between thousand-acre growers and the sales outlets that service worldwide retail markets for fruits and vegetables, including, of course, seedless watermelons.

In the May 7 1988, issue of *The Packer*, Tom Zind tells us that Sun World

> . . . appears poised to wake up what has been a rather "sleepy" industry. . . . The combination of advances in plant and growing technology, consumer interest in both convenient foods and unusual produce, and backing by large produce marketers, have brought the seedless [watermelon] varieties front and center. . . . The company's venture into seedless began two years ago with only small acreage and experimental marketing. This year, however, the company is going full bore, targeting promotions and in store demonstrations in major markets throughout the shipping season of April through October, Sun World, spokesman David Margoulis said. We introduced the Flame Seedless grape 10 years ago, and because of its success there were thousands of acres planted. . . . We think seedless watermelons will be as big an item in the melon category as Flame Seedless grapes were in the grape area. . . . There's a tremendous application for it in food service, but to date we've had a fantastic demand at retail as well.

Zind's article also tells us,

> American Seedless Corp., which merged with Sun World, once was the primary supplier of seeds for the seedless watermelon. That company's seeds now are sold only to growers for the Sun World–American Seedless joint venture, American Seedless. But as interest in seedless grows, more seed companies are working on coming up with a seedless variety.

There is something deeply disturbing here. "Seedless Poised For Sweet Success" is the title of Zind's article. There is no mention of O. J.'s failures, or those of anyone else, for that matter. Readers would be unaware of the soul-testing frustrations that preceded success. Nor would they see the story of the seedless watermelon as a real example of opportunities in America throughout the twentieth century. In O. J.'s words, "This is a marvelous example of a person who is allowed to either succeed or fail and I did a lot of failing. But I learned from my failures."

Chapter 10
From Son to Father to Grandfather in the 20th Century

THE INTIMACIES OF family relationships deserve, in fact, demand being shielded from public display. At the same time, to exclude his role as son, father, and grandfather would make the life story of O. J. Eigsti incomplete. Fortunately, to treat this phase of O. J.'s life generically avoids sins of both omission and commission. Equally fortunate, the work of William Strauss and Neil Howe provides a wider view. Their book *Generations: The History of America's Future* enables us to see the satisfactions and frustrations, the ups and downs, challenges, opportunities, responsibilities, everything that comes with being son, father, grandfather not as personal attributes, strengths or shortcomings, but as continually evolving interaction between a person and his or her society.

In effect, Strauss and Howe's analyses of the generations that will have played out the major portions of their lives during the twentieth century provide wonderful perspective. In turn, to focus on the life of one person not only fleshes out this perspective, but invites each of us to enter their analyses, to reexamine what we have done and where we have been during the twentieth century. Perhaps more pragmatic than philosophical, we can set goals as we look toward the new millennium.

O. J.'s father was a member of the Missionary generation (born between 1860 and 1882). The following phrases are selected from among those used by Strauss and Howe to characterize this generation:

- Indulgently raised and educated.
- Preachy student leaders.

- Rebellious career women.
- Moral pathfinders to whom any opinion was a religion once they decided it was right.
- Reveling in the independence of their new automobile.
- A fierce desire to make the world perfect.
- Vindication of social good over social evil.
- Odd detachment from the material realities of life.
- Lifelong quest for world reform.
- Youthful indulgence, coming-of-age fury, rising-adult introspection, and midlife pomposity and intolerance.

It isn't easy to observe these generation characterizations expressed in the relationships between O. J. and his parents. It is easier to question how accurately these generalizations apply. As previously brought out, the unique character of people in rural America may not have received adequate attention in the analyses conducted by Strauss and Howe. They may have not taken into account that in certain communities, religious tenets or ethnic traditions outweigh more generally applicable societal forces.

"Reveling in the independence of their new automobile" provides a case in point. The Eigstis were among the first in their community to own an automobile. O. J.'s father owned two 160-acre farms, one serving as the family home and the other operated by relatives in Iowa. Visiting with these relatives, together with discussing financial and management affairs, required an annual automobile trip between the two farms. The ritual of this trip, however, does not suggest "reveling in independence." There was much planning. A farm hand was the designated driver. O. J. describes it:

> *In 1911 we bought the first automobile and used it to go from Morton, Illinois, to Wayland, Iowa. The first trip that I definitely remember is our September journey in 1914. It was to attend the golden wedding anniversary of my grandparents, the Sebastian Gerigs. Pete Warner, our hired man, was the driver. Emery and father sat in front, Mother, Minnie and Mahlon in the back seat. Willis and I sat on orange crates between the front and back seat. To round out the total passengers, Clayton was just a bit over one year old.*
>
> *We started out at dawn and if all went well, we expected to arrive in Wayland by evening. Of course, the roads were dirt and only a bit of pavement in such cities as Peoria and Knoxville, Galesburg and Monmouth. We crossed the river at Oquakua on a ferry. It was exciting to cross the wide Mississippi. One time the motor that pulled the ferry stalled and we started drifting down stream. But they got it started and we reached the other side.*

On this particular trip we had one of those heavy fall rains when we got to Knoxville. They were just laying brick outside the city and we got stuck in the mud. The car broke down with a broken axle. A neighbor pulled us out and we had dinner with them. After dinner, he loaded us in his wagon and we went to the railroad station to catch the train the rest of the way to Wayland. We arrived before evening.

There is more here than O. J. tells us. The Eigstis' first automobile was a Moline Dreadnought. One wonders why a name that suggests a battleship might add sales appeal to an automobile. It certainly did not look like a current model. There were no doors to the front seat. The top folded back and was raised only during stormy weather. The windows were covered with isinglass, the thin, reasonably transparent sheets of the mineral mica, which was the equivalent of nature's first plastic. And running boards on both sides carried luggage or whatever had to be transported. There was no electrical switch to start the engine running. For this, a crank had to be turned by hand to start the pistons doing in the cylinders what explosions of fuel would do later.

The nation's roads weren't ready to pamper automobiles—thus, the broken axle caused by bad road conditions. And can you imagine a modern motorist having his "dream machine" pulled into a farmyard by a team of horses? Or the farm family putting up the designated driver until a replacement axle could be obtained and installed? This happened, but Pete Warner didn't finish the trip to Wayland. He returned the repaired Moline Dreadnought to Morton and the Eigstis came home by train.

There are several points to be made here. Although we make much of how things during O. J.'s early years differed from today, they worked, they got the job done. Progress has been defined more in terms of improvement than in totally new ways of doing things. This one travel episode reminds us that railways enabled people to travel from place to place, even between obscure places. It is interesting that O. J. simply does not remember how the family obtained food during the ten- to twelve-hour journey. Nor does he remember fussing over other discomforts. An average of three or four flat tires may have had something to do with this. Stops along the way took time to remove the damaged tire, find the puncture, scratch the rubber, apply cement, peel away the protective cover, apply the patch, and install and pump up the tire.

If there is a point to be made here by contrasting the discomforts of O. J.'s early years with modern-day pampering, it is that the way survival problems were solved did not preclude solutions to other problems. Flat tires or primitive facilities along the way did not prevent cousins from getting to know each other, family traditions from being kept alive, and relationships from being continued.

Recalling the advances in transportation that have been a part of the Eigsti life story, O. J. continued to tell about his introduction to "the first wireless—radio, that is, in 1917 or '18."

> *Lester Litwiler was an electronic genius. He built crystal sets in his room that could receive music from far away. We used head sets or earphones and the first reception I remember came from Pittsburgh. I believe the station was WOW and this was about 1917. We would spend hours tinkering and listening. Then he built a set we took home where we spent a lot of time listening to music and talking.*
>
> *Another big step in hearing about telecommunications came when I met Dr. Tom Goldsmith at Cold Spring Harbor. Tom had just completed his degree in physics and it was related to television. This was 1937 and he had an oscilloscope that recorded waves. He used this device to try to explain to me that some day we would see actual pictures on the screen. . . . Tom went to New Jersey to work for one of the first companies to make television sets. I can say I lived through a period of great development in communications. I wonder where that first crystal set I listened to is today.*

Another incident cuts across and, in effect, negates several of the previously cited generational characterizations. Christmas in the Eigsti home was well organized. Gift giving was limited to drawing one name from the hat. The grandchildren remember books being the predominant gift, many of which were read through before the day was over. At the proper moment, everyone would take their places in the "parlor" and the obvious head of the household would ceremoniously remove a pack of envelopes from his suit-jacket pocket and present one to each of his sons and daughters. Each recipient knew, of course, that each envelope contained a crisp, new $100 bill. It is significant that this ritual continued even when the recipients were well into middle age.

To cite the "rebellious career women" characterization brings additional anomalies. Only church-related activities competed with home and family responsibilities for O. J.'s mother's time and energy. She was true to the Mennonite faith. Attendance at a midweek and two Sunday services was taken for granted. She is remembered as never being seen without the head covering that meant so much to her.

Nor does the "odd detachment from the material realities of life" characterization apply to the lifestyle in which O. J. grew up. Making money received much attention. Farm produce of all kinds was sold both wholesale and retail. Home-butchered meat found a ready market both locally and as far away as Chicago. Squabs were a specialty. At times the Eigstis would butcher their neighbors' animals and receive as pay the makings of their own sausage. O. J.'s father played his version of the futures market by storing his grain in the local grain elevator and calculating the time to sell so as to receive the best price. This led to becoming friends with other like-minded financiers, particularly at the Morton State Bank, where he often would be asked to visit farms whose owners were seeking loans. It soon became known that only well-tended farm property would merit his approval. As O. J. remembers this,

> *My father was in charge of approving farm loans and at his funeral, the president of the Morton State Bank came to me and said, "We are tremendously grateful to your father. If it hadn't been for your father's judgment on loans to farms, our bank would never have survived." He got terribly criticized. What happened was they got so mad at his bank that they formed another bank called the Farmers State Bank down the street. So my father would say, "we can't make that loan but you may want to try down the street." And they were one of the first to close up, to go bust when the Depression came.*

When everything is taken into account, relationships between O. J. and his father were pretty much a rich distillate of role model catalyzed by empathy, often energized by divergence of opinion. Lessons were taught, sometimes to be taken advantage of, other times to be contradicted, but always portraying life as best lived at full speed.

The GI generation, those born between 1901 and 1924, includes O. J. and Agnes Eigsti. Phrases selected to characterize this generation include:

- Truly "special" kids.
- A coming-of-age slingshot (World War II).
- An enduring sense of civic virtue and entitlement.
- Enormous economic and political power.
- Unstoppable energy.
- In Kennedy's words, to "bear any burden, pay any price" to accomplish whatever goal it sets.
- Godlike in its collective, world-bending power.
- Adept in its aptitude for science and engineering.
- Father-worshipping and heavily male-fixated.
- The most conformist culture.
- Beneficiary of an unmatched flow of payments and other kindnesses from people older and younger than they.

Logic suggests that who the members of the GI generation became, their motivations and values, was influenced by those of the Missionary generation. Which, of course, is another way to say that Agnes and O. J.'s motivations and values were influenced by those of their parents. But we are not ready to say that such influence reached the proportions and had the character of the generic influence cited by Strauss and Howe. It seems to be overblown to say that "by degrees, these young men and women entirely reshaped American values and culture. While Christian socialists preached the Social Gospel to urban immigrants, Billy Sunday preached 'fundamentalism' to an emerging rural Bible Belt. While poet-guitarist Carl Sandburg, 'Ragtime' Scott Joplin, and 'Father of the Blues' W. C. Handy enlivened Greenwich Village."

Sons Karl and Nick Eigsti were born during the transition between the generations, the Lost generation, those born between 1925 and 1942, and the Baby Boom, or Boomers, those born between 1943 and 1960. Strauss and Howe identify the Lost generation as "the generation stuffings of a sandwich between the get-it-done G. I. and the self-absorbed Boom." On this basis, we would see the Eigstis' sons' generation infrastructure as characterized by such phrases as:

- A lack of surefootedness;
- A keen sense of how and why humans fall short of grand civic plans or ideal moral standards; Badly checkered family lives;
- Substance abuse moved past euphoria to social damage; A mixture of high self-esteem and selective self-indulgence;
- Better philosophers than scientists; better preachers than builders;
- More independent of social bonds, yet more open to emotional isolation and economic insecurity.
- Measure themselves subjectively by the spiritual strength they see within.

It isn't easy to trace these attributes of the Eigsti sons' generation to the values and motivations being put into practice by the reigning G. I. generation. The best we can do is cite some of the more significant characteristics that showed through while the G. I. generation was "tending the store."

> Returning war heroes brought a mature, no nonsense attitude wherever they went—to campuses, to workplaces, to politics. . . . 40-year olds of both political parties agreed that big government and big business could "pitch in" and work together just fine. But [the] Vietnam [War] helped trigger an angry "generation gap" between G. I. parents and coming-of-age Boomers, along with urban riots, a crime wave, substance abuse, eroticism, and ideological passion—in short everything hateful to the G.I. life mission. . . . By the late 1960s, whatever G.I.s tried (whether "Model Cities" or "guns and butter" economics) began to sputter. . . . men came under constant attack from juniors and, increasingly, women. If the times were euphoric for youth, they were anything but for G.I.s: "Everything seemed to come unhinged" (James McGregor Burns); "Something has gone sour, in teaching and learning" (George Wald); "I use the phrase soberly: The nation disintegrates."

The Eigsti grandchildren are members of either the Thirteenth or Millennial generations (also known as Generation X). These generations are characterized by such phrases as:

- Feel that the real world is gearing up to punish them down the road.
- More than half are immigrants or children of immigrants.
- Seen by their elders as being frenetic, physical, slippery, shocking on the outside, unknowable on the inside, driven more by appetites than by ideas.
- See themselves as pragmatic, quick, sharp-eyed, able to step outside themselves to understand the game of life as it really gets played.
- Early access to self-expression and independence stripped them of much of the pleasure of discovery and rebellion.
- Matter-of-fact approach to sexuality.
- Do what they feel they must.
- Proud of their ability to poke through the hype and the detail.
- Delight in the demeaning images of youth crafted by the electronic media.

Relationships between grandparents and grandchildren are uniquely difficult to put into words. Continuing to quote Strauss and Howe:

> The G.I.s' material well-being has enabled them to continue in their accustomed roles as trustees of wealth, givers of gifts, acceptors of collect phone calls—and providers (or backstops) for their extended families. . . . But if Silent and Boom children frequently ask for G.I. assistance with economic problems, they do not seek elder help nearly so often on questions of values or basic life direction. . .. Whatever wisdom G.I.s do have to offer, they lack the pulpit to proclaim it, owing to their early retirement and the recent Boom takeover of positions in the media. . . . Yet their final ledger will also include colossal debits, unprecedented public and private liabilities, exported assets, depleted resources, harms to the global environment.

There are features in the life story of O. J. Eigsti that are generated both within and beyond those that come with being a scientist, agriculturist, teacher—features that may well have had the most to do with determining his role as son, father, and grandfather. There is more here than apologies for short-term impatience, intolerance, and similar personal traits. It is not a matter of "having gotten up from the wrong side of the bed" on a particular day. On the contrary, it is a case of one's image as seen by others, both at home and away from home, being influenced by both the actual and side-effect experiences that come with earning a livelihood. For O. J., there is added complexity that these influences have come from three diverse livelihoods.

There seems to be something about farming, something about producing corn or watermelons, hogs or chickens, cotton or chinchillas, that characterizes farm people, at least generically. Farmers are often thought of as being free to do things their own way, to make their own decisions, to set goals, to take chances—in short, to enjoy the full benefits that come from living in a free society. But farming has undergone tremendous changes during the century that is represented by the life of O. J. Eigsti. There is a diminishing fraction of the U.S. public who have known life on the farm—a sharp decrease in those in other jobs or professions who carry with them the attitudes, values, and beliefs that characterize farm people. At the same time, there has been extensive "cross-fertilization" between farm people and their city cousins. Few of the distinguishing traits in either group seem to be strictly dominant or recessive.

Other authors have written interesting and worthwhile books that document the ups, downs, satisfactions, setbacks, seasons of drudgery, and moments of deep emotional reward that characterize farm life as it was lived during the early part of the twentieth century. The more courageous of these authors trace the origins of lifelong personal traits to interactions with societal changes.

I indulge here in only one example. O. J. has never been at ease with blue jeans and bib overalls being seen as stylish and to be worn at all occasions. To O. J., overalls are a uniform to be earned, the "purple heart" awarded after surviving the endless hours of physically demanding, often filthy, dirty, dusty drudgery of animal husbandry and crop management. Overalls were worn by people trying desperately to see their crops through days when the rains would never come and keeping their livestock alive when even the proverbial grasshoppers had to carry their own lunch-buckets.

Which brings me to a large portion of what it has meant for O. J. to grow up in the twentieth century that we can do no more than scarcely recognize. Changes in style and fashion, how we dress up or dress down, the messages we are permitted to broadcast by choice of costume, are but one of several elements of our society that have changed vastly during the century. Especially so for someone whose boyhood was pretty much prescribed by the ethical values and moral standards proclaimed by the Mennonite faith and the farm milieu.

Attitudes toward sex and sexuality have undergone changes that could never have been foreseen. Sex and all of its attributes left the bedroom and the barnyard. Discussions of sexual taboos moved far beyond biblical no-nos to become media topics that are sure to gain a paying audience in any medium. Strict adherence to the belief of working only six days of each week has evolved into practically no restrictions on bar code readers or robotics assembly lines. Greed has become good. Drug addiction has moved from being the result of patent medicines laced with narcotics or the faulty judgment of "needle-happy" family doctors, to mind-altering substances treated in the media as "recreational" substances that may lead to chemical dependence.

O. J. was not unaware of these societal changes and their effects on his own life, family, friends, neighbors, colleagues and business associates, and on the nation in general. On the contrary, he admitted to a kind of impotence, to the futility of jousting with the windmills of his day, and reserved his time, energy, and enthusiasm for the pursuit of what he believed would be achievable goals. More important, he saw the media's treatment of societal defects and dysfunctions as being overblown, done for effect, and as self-serving efforts on the part of the few. He was always ready to give each person he met, each student in his university classes, each colleague or competitor in business, a fair chance to radiate and practice her or his own brand of personal and professional integrity.

All of which is to say that there are times when we can best understand O. J.'s role as son, father, and grandfather by remembering that he was bred and born on the farm, schooled as a farm kid, went to church with rural neighbors, and received his advanced degrees at a land-grant university. He earned his livelihood as farmer, scientist, and teacher, but was always required to adjust to complex forces that cut across all facets of our constantly changing society.

Nor were these forces feeble (in any sense of the word) during the twentieth century. O. J.'s early boyhood was dominated by World War I. His early manhood was dominated by the Great Depression. Then came World War II—few events in American history cut deeper into as large a fraction of the lives of her people. And what period in our history has known greater resurgence in science and technology than the aftermath of World War II? No greater tampering with a nation's ethical and political infrastructure has been seen than that which swept across America during the 1960s and '70s? For the nearly three decades of what for many would be retirement years, O. J. has continued to adjust to a society that bears little congruency with that of his Mennonite family heritage and rural upbringing.

Borrowing insights from Carl Hamilton and his book *In No Time at All* (1974), Eigsti's family lived through the "'slump in farm prices' after World War I until World War II." He scratched out an education while "FDR and the New Deal turned things around but only to the extent that farmers could 'hang on.'" But that didn't keep O. J. from seeing and feeling the effects of "10-cent corn, 3-cent hogs, loss of farms, [and] foreclosures." He was no stranger to agitation for "farm relief" or how opinions differed about government "tinkering."

No one needs to explain to him why neighbors at "penny auctions" would bid pennies for farm machinery at sheriff's sales and then return the items to the original owner. Or why "the sullen crowd, some armed with pitchforks and ax handles, quickly gave the word to anyone not playing the game." Then there was the drought of the '30s. "Day after day of searing temperatures and hot, southwest winds. The corn would roll and fire. Chickens would seek the shade and flop down, wings extended, beaks gaping open. All living things sought shade. There would be lightning at night and new hope. But it was heat lightning! Each day and each hour of the day you would scan the skies for clouds. They didn't come and at breakfast someone would predict 'another scorcher.' He was right."

Nor did O. J. escape the terribly depressing and environmentally destructive dust storms that were an aftermath of the drought. By 1936 he had left Illinois for a position at Long Island, but news from relatives in Illinois and Iowa weighed heavily on his mind. Again quoting Hamilton, "the sky would darken and cars would drive with their lights on. Everything you ate, touched, or wore felt gritty, even the sheets on your bed. It was not just a good thick layer of dust throughout every room in the house and in every cupboard and closet; there would be drifts of dirt around every loose-fitting door and window." The ultimate financial and sociological impact on life in America's "breadbasket" has been well and honestly documented.

Stepping aside from the generic approach, without doubt O. J. Eigsti's role as son, father, and grandfather has included the ups and downs, the plusses and minuses, the good days and the bad, decisions both wise and foolhardy—decisions that came with a life being spent during the twentieth century. However, hardships and horror stories rarely show up in O. J.'s repertory of conversational topics. Setbacks are to be regretted, that's for sure. But little is gained by seeing them as punishment for faulty decisions. More than this we leave to the memories of those to whom he has been son, father, and grandfather.

Another friend, Harold Smock, titles his autobiography, *Pebbles in the Stream*. The metaphor serves well to close this chapter. The life of O. J. Eigsti, or of any of us for that matter, can be seen as the momentary splash caused by the dropping of one pebble from a universe of pebbles into a flowing stream. For sure, the environments of all nearby pebbles are disturbed, for the better in some cases, for the worse in others. Even after this new pebble comes to rest, it will forever affect the overall flow of the stream in some small way. And equally so, the environment of this pebble will be affected by the splash of new pebbles arriving ever after.

Chapter 11

O. J. Eigsti as a 20th Century Scientist

The life of O. J. Eigsti gives us a wonderful opportunity to think about how science has evolved throughout the twentieth century. He grew up with the urge to understand the natural world, to feel at one with and to make nature and humanity mutually beneficial. As we have seen, there was little formal education in science in his elementary school. Physics was the only science course offered at his high school. In a very real sense, O. J.'s introduction to science was by way of growing things, tending livestock, adjusting to daily and seasonal environmental changes, operating and repairing farm equipment in which physical principles were applied, and taking in stride both the helpful and harmful effects of weather and climate.

At the college level, O. J.'s career interests centered more on education and the social sciences than on natural science. However, when no jobs were available in these fields, it was the uniquely rewarding courses in the life sciences, particularly botany and genetics, that nudged him in this direction in graduate school. When his master's and doctoral theses were approved and he passed the traditional oral examinations administered by committees of gatekeepers, he entered the official realm of science with all of its traditions, opportunities, and responsibilities.

The prevailing character of the life sciences at the time O. J. entered the profession is suggested by the titles of his master's and doctoral theses: "A Contribution to the Life History of Reseda odorata L." and "A Morphological and Cytological Investigation of Certain Members of the Resedacaceae." Describing the life histories of representative plants and animals commanded much attention. Identifying relationships was limited pretty much to providing a means for classification, that is, putting individual species into their proper taxonomic division. Some scientists, however, were giving special attention to events or circumstances that suggested relationships between species or pointed toward a common origin. These interests led to the establishment of fundamental principles that both grew from and helped account for key features that recurred in the life cycles of different species. And there were scientists who were most curious when their observations seemed to contradict what the application of established principles would have led them to expect.

From the beginning, there was no question that O. J. would specialize in the life sciences, particularly the plant sciences. Purely physical phenomena, those that involve energy and its transformations in the absence of the life entity, never captured his interest. His life as a botanist pretty much parallels that of science itself. His first publications reported observations of one or another of nature's creations. In time, his attention was drawn to what appeared to him to be a discrepancy, something unusual happening in nature, something that caused an unsettling feeling until its action could be seen within the bounds of an orderly, predictable nature. Finally, once he believed he understood selected events in nature, he set out to put his newfound knowledge to use.

In less complex language, O. J. began as a "pure" scientist and advanced to practicing a science-related technology. Along the way, however, he always took special pride in those situations where a solution to a technical problem became possible only by reverting to the role of "pure" scientist and ferreting out just which of nature's actions were the cause of that problem.

Eigsti's formal résumé is a literal paper trail of his life as a scientist, particularly the first twelve of the twenty-nine years he spent as a university professor. His published work includes four publications that appeared in 1938, two in 1939, six in 1940, seven in 1941, four in 1942, three in 1943, and one in 1946. For now, I only mention that colchicine is included in the titles of five of the eight publications that appeared between 1947 and 1961. Polyploidy, equally involved in the science and technology of seedless watermelons, appears in the other three titles.

The abstract of an article that appeared in 1937 in the *American Journal of Botany* is of special interest. The title is "A Comparative Study of Male Gametophytes in Tradescantia reflexa Raf. and Lilium Canadense L." Two sample sentences are: "Chromosomes in generative nuclear divisions in Lilium separate from male nuclei from prophase without equatorial plate organization, whereas the same in Tradescandia form male nuclei from prophase without equatorial plate organization. This method yields data valuable for studies of chromosome structure and the interpretation of polyploidy and sterility incompatibilities."

When O. J. compares observations that involve two species, he advances one step above the beginning level of the pursuit of science. References to polyploidy and sterility point toward broad principles. Although the seedless watermelon was many years in the future, it is significant that both of these principles play key roles in the technology of seedless watermelon production.

Evidence of another step forward—this one truly a giant step—exists in an article titled "A Cytological Study of Colchicine Effects in the Induction of Polyploidy in Plants," published in the *Proceedings of the National Academy of Science* in 1939, based on an investigation completed by Eigsti in February 1937. Quoting the article,

> This initial cytological study using colchicine became centered around the following points: (1) the effect of colchicine upon individual embryonic cells rather than entire tissues; (2) the degree of polyploidy in a cytologically changed nucleus and the relation of the change to the method of treatment; (3) observable cytological changes due to the effects of colchicine upon the mitotic process; (4) the probable effectiveness of colchicine as a means of producing hereditary changes.

This article is a significant step because of its contents and is a giant step because of its methodology. It goes beyond merely observing events and circumstances as they occur in nature. The event under study is brought into the laboratory and caused to occur under conditions that the investigator can tamper with, expecting to observe elements of the event that have not yet been revealed. In this specific example, the total environment in which the event is to occur is reduced to the tips of roots appearing on an onion submerged in a water solution of colchicine.

Illustration of another step along the pursuit of science appears in "The Effects of Colchicine upon the Division of the Generative Cell in Polygonatum, Tradescantia, and Lilium," an article published in the July 1940 issue of the *American Journal of Botany*. Illustrated here is the practice of bringing together the investigator's previously acquired knowledge; adding, perhaps, related knowledge gleaned from the work of other investigators; introducing a new hypothesis or new environmental conditions; and observing the effects, if any, on the event under investigation. Quoting the article, "it appears that further experimental work with pollen tubes presents the unique possibility of combining two objectives, viz., that of studying the nuclear and cytoplasmic changes brought about by the action of colchicine upon mitosis in gametophytic cells and that of determining the mechanism of division of the generative cell in pollen tubes."

We now skip ahead some seventeen years along the paper trail that marks O. J.'s journey in the pursuit of science to "Induced Polyploidy," an article first published in 1957 in the *American Journal of Botany*, and in 1958 in *Fifty Years of Botany, Golden Jubilee Volume of the Botanical Society of America* (edited by W. C. Steere, published by McGraw-Hill). The agenda underlying this paper is revealed in the opening sentence: "More food and fiber is produced from polyploids than from the related species that are diploid." Furthermore, "the present human population depends upon the greater increases in annual production that come from the polyploids against the lesser quantity that would be available if only diploids were under cultivation."

To back up this claim, O. J. tells us that "our daily bread is made almost entirely from the flour obtained from hexaploid and tetraploid species of Triticum. Cotton in largest supply is produced by the tetraploid rather than the diploid species. . . . About one-half the species of flowering plants are polyploid. . . . The superiority of ornamentals and horticulturals that are known as natural polyploids keeps alive keen interest in polyploidy as a practical means for producing new and better flowers and fruits."

This quotation particularly justifies saying that O. J. has arrived at that step in the science enterprise where the scientist is no longer content to pursue knowledge for its own sake. There comes the urge to contribute directly to humanity's well-being and be rewarded for doing so. O. J. lets his readers know that the hundreds of investigators—who have not only taken the mystery out of polyploidy, but have learned to manage and, to a degree, control this bit of nature—deserve recognition. Of course, colchicine deserves equal credit for its role as an essential "laboratory assistant." Nor should his audience be surprised to have O. J. proclaim that polyploidy has enabled watermelons to join bananas, grapefruit, grapes, pineapples, and oranges in sharing the advantages of being seedless.

We come now to a step in O. J.'s voyage through the realm of science that is difficult to discuss, fraught with temptations to invade the privacy of personal correspondence or read between the lines in search of documentation of things that can be only dimly seen. There is another problem: much of my conversation with O. J. throughout these many years was no more than informal sharing of mutual interests—gossip, if you will—about friends and strangers whose work, beliefs, attitudes, ambitions, and goals crossed paths with ours. With a tape recorder between us and knowing that creating this life story is under way, there should be no need to censor, no need to apologize for whatever shows up on the recording. In fact, only by "playing back" O. J.'s reactions as they were expressed can I convey certain features of who he is and what he stands for. I nevertheless feel a need for apology.

The following conversation began with my reading to O. J. a working draft of the portion of the manuscript reporting on some Chinese articles about seedless watermelons and on the letter from Dr. Bill Rhodes—the conversation that prompted the apologetic note. Quoting Dr. Rhodes's letter, "Japan has developed and implemented the technology to graft large numbers of hybrid seedlings on disease resistant root stocks of the same or different species."
O. J. responds,

> *I've done that grafting myself. Total failure. You can take these great big dwarf gourds, big gourds, I've forgotten the name. I've done this grafting and you can get them to grow, but it's not practical. Very impractical. [It works as an intermediate step in tissue culture,] but then the thing to do is to get genetic resistance like I've got out here with my crosses between Calhoun Gray and wilt-resistant Congo. These melons are growing where melons have been grown every year since 1949. You don't need to graft on cucumbers when you have these disease-resistant, wilt-resistant lines of tetraploids.*

Returning to Dr. Rhodes's letter, "This technology will probably play a role in the mass propagation of seedless watermelons. "
O. J. says,

> *That's a bust! Absolutely a total bust.*

The letter continues, "The tissue culture generation of diverse tetraploids which can be used as parents of new triploids is routine. Colchicine will probably be replaced by common herbicides as the chemical inducer of tetraploids."
O. J. responds,

> *Total bust! Absolutely failure. That's just a total failure. That's just nonsense.*

The letter continues, "Once commercial cloning of seedless watermelons becomes commonplace, it will not be necessary to establish homozygosity in the parents and individual hybrids can be increased to staggering proportions for growers."
O. J. remarks,

> *All of this has been totally . . . he's not up to date on this.*

Rhodes also tells of taking lessons from the hybrid corn industry and using the male sterility phenomenon. O. J. responds,

> *That's your male sterile tetraploid and we've got that from China and it's a total bust. You know what's wrong with the male sterile watermelon? It's also pollen sterile. This male sterile diploid*

takes the other half out of getting triploid seed to produce fruit. Oh, this male sterile line was going to revolutionize the diploid hybrid seed production in watermelons. But what they discovered was that it was not only male sterile it was pollen sterile. You couldn't reproduce it. And the Chinese think they're going to develop seedless watermelons by this male sterile bit and they're not going to. It's not cytoplasmic sterile as in corn. It's chromosomal. See, corn is cytoplasmic male sterile.

Rhodes shares with O. J. the same optimistic view of the future because seedless watermelons have great potential in the United States and around the world. Production will continue to increase. Our export market should grow because seedless melons can be shipped abroad. O. J. agrees,

Oh yes. We're shipping by air and by ship, we're shipping seedless melons to Japan right now.

Continuing to quote the Rhodes letter, "The tenacity and ingenuity of Eigsti in growing the seedless watermelon into reality in the United States is understood and appreciated by all of the next generation of watermelon scientists. "
O. J. says,

That's exactly my contribution. Oh that's good. That's exactly right!

This is confirmed in "O. J. Eigsti and His Love of the Seedless Watermelon," which appeared in *The Cucurbit Network News* for spring 1996. "In spite of the many obstacles, O. J. Eigsti persevered in his research on and marketing of seedless watermelons. . . . His steadfast determination over the past 60 years to produce high-quality seedless watermelon has turned a cucurbitologist's dream into a tasty, seed-free reality."

What does all of this really tell us about O. J. Eigsti and his role as a scientist throughout the twentieth century? Is he representative of who scientists are, their values, traditions, and the ways they work? Is he out of character when he seems to be less than open-minded, unwilling to explore new ideas, new ways of doing things, hesitant to give up familiar beliefs and practices and at least entertain those of a new generation whether or not they are "user friendly?"

To be fair, we cannot answer these questions without taking into account the full breadth of his experiences, particularly his interactions with other people. Remember that he went directly from graduate school to rubbing elbows with colleagues who would later become Nobel Prize laureates. For several decades, he moved comfortably among other scientists, attended their meetings, listened to and presented scholarly papers, and participated in seminars—not just in the United States. His first business partner lived in Japan, his coauthor lived in Belgium.

Dr. Xingping Zhang is a good example of O. J.'s interactions with other people. Citing personal correspondence from him,

> I fully respect Dr. Eigsti's important role in exploring and commercializing seedless watermelon in the United States. I met Dr. Eigsti the first time in South Padre Island, Texas, in November, 1994, during the Cucurbitaceae '94 conference. I spent a lot of time talking with him about seedless watermelon, genetics, and breeding. I fully appreciate his attention to a young scientist like me. I met him again during ASHS 92nd Annual Conference in Montreal in 1995. Again, our time was spent talking about seedless watermelon.

Both as a former colleague of Dr. Rhodes at Clemson University and as a scientist in his native country, Dr. Zhang adds to information already cited.

> The use of tissue culture in micropropagation of triploid seedless watermelon was initiated in China in early '70s due to the difficulties in producing high quality triploid seed, and the difficulties of growing vigorous seedlings from the triploid seed. The first commercial size field was grown from tissue culture propagated triploid watermelon in Beijing in '84 but that was not repeated since then. Tissue culture can produce high quality seedlings and the tissue culture seedlings do produce good quality fruits. However, that research was not commercialized because the seedlings produced by tissue culture cost much more than seed. . . . The tissue culture method, I believe, is more efficient and feasible for developing new tetraploids for the breeders who have access to tissue culture facilities. I have demonstrated that the time needed for developing a new tetraploid can be shortened from 6 (or longer), years to 2 years if the tissue culture technique is properly used.
>
> The grafting technique has been used for diploid watermelons in countries like Japan, Korea, China, and Spain. The main reasons for using this expensive process are to enable watermelons to grow in the same soil without long period rotation and or under low soil temperature, and to enable watermelon plants to produce fruits over a longer period of time. . . . The main drawback is the cost; the Japanese even made a robot for grafting but it was too expensive to use.
>
> The most common chemical used to double chromosome numbers is colchicine. A lot of experiments have been done in watermelon since Dr. Eigsti's initial work. All of the commercial

tetraploids were developed by using colchicine. Some herbicides can effectively double watermelon chromosomes. I consider this only another option besides colchicine. And as for the male sterile mutant, it can be very beneficial for reducing the cost of triploid seed production.

Dr. Zhang's letter illustrates how O. J. has contributed to and benefited from interactions with certain colleagues. Now O. J. tells us about moving equally beneficially in a different setting.

I saw these melons at our grocery store and I knew they were from our Tri-X313 seed. They were beautiful. Everybody was bragging about these seedless melons. I went to the produce man and [asked], who in the world is producing these melons. He said, "an Amishman up there between Shipshewana and Middlebury" [Indiana] but he didn't know his name. So I went to the cheese house at Middlebury and found this man with a beard and asked him who is producing these good seedless watermelons. He said, "I know who that is. That's Goliath, David Goliath Bontregar, he lives straight across that field."

So I came home and told Amsie, Amsie Miller our Amish farmer, that I had just visited with Bontregar and he said, "Oh you mean Goliath." And you know he was barefooted and I've never seen a man with such big feet and such big hands and he's a giant. Amsie said, "That's why we call him Goliath." And he's very much Amish, and organic. Has eight children and he's forty-five years old. And he sure knows how to raise seedless melons. And when to pick them. I had a wonderful visit with Goliath!

The point being made here is that O. J. is not restrained by stereotypical notions of what it means to be a scientist. He has his own yardstick for measuring the worth of another person, what he or she claims to have achieved or hopes to achieve; and, particularly, the chances that what they promise will be fulfilled. The trauma from promises, his and those from others, that could never be fulfilled lives on. Better to judge other people, particularly scientists, only by what they are doing rather than by what they promise to do or want to become.

For now, there is something to say about the spirit that shows through O. J.'s comments while listening to Dr. Rhodes's letter. Remember that this was private conversation and is being included only because he speaks to a significant facet of O. J. There is a bit of the "loose cannon" syndrome in his persona. Occasionally, it is difficult to understand where what he is saying is coming from. It is as though his intellect is momentarily overloaded, overcharged with an urge to protect what he believes in. During such moments it would be unfair to take everything he says at face value. It would be better if we balance what is being said against other things we know O. J. believes in.

Consider, for example, the following paragraph that appear in a paper titled, "Genetics, Fundamental and Applied."

> Today many geneticists are working in a highly theoretical field of genes, chromosomes, and molecules. Frequently their work is very difficult to interpret and the ultimate applications are not known. The very nature of their work is such that only rigid adherence to a scientific method can be their guide. Judging from past experience we can be assured that when some of the solutions come to our present day problems of securing better crop plants or securing relief from diseases that vex the best scientist, these solutions will come from those scientists who have studied the genes, chromosomes, and molecules. In light of what has been said, it is better if we do not make the common mistake of calling the man who studies the eclipse a stargazer and praise the man who designs a television set the miracle man. Both are necessary. I would not call the biologist a bug chaser and point to the sanitary engineer as a great humanitarian because he knows how to prevent the spread of disease. Do not classify geneticists as too academic and hail the man who brings the bag of hybrid seed corn to your farm as the plant wizard. Do not mistakenly call the virologist a wasteful researcher and the man who gives you polio vaccine the great healer.

I now return to "the theory of gemetophytic selection with special reference to tetraploids" to flesh out a facet of O. J.'s life as a scientist. As is often true for the conception of scientific theories, it begins with a relatively simple observation, specifically, that, today, the triploid Tri-X313 seedless watermelon is more uniform, with vigorous vine growth, better quality, and better adaptation to varying environmental conditions, than when it was first tested in 1957. O. J. traces these changes that show up in the triploid hybrid to the tetraploid parent, specifically, the #4008 tetraploid. This parent seed line, when tested in 1991, showed greater vigor, more uniformity, and better productivity than it did in 1949.

O. J. sees this as resulting from natural selection in the open-pollination fields acting throughout many generations. Genes that program for poor characteristics are weeded out when growers retain those melons that measure up to their expectations. O. J.'s explanations of how this happens, however, is much more academic, drawing more on his armory of familiarity with basic science. I have already mentioned O. J.'s introduction to the key scientific concept of gemetophytic selection. He sees the life history of the watermelon plant as retaining a brief remnant of one of the original "parents" of all seed-bearing plants. Brief, yes, but existing long enough to allow nature's relentless selection processes to eliminate the weak and favor the strong.

In effect, the enormous number of pollen grains that appear on the anthers of watermelon flowers represent a separate, short-lived generation, but their tremendously large number is sure to suggest the notion of survival of the fittest, the fittest in this case being the ones that unite with the many, many times fewer ovules. On this basis, nature alone does what no technology can do; slowly, yes, and in its own way. It is O. J.'s faith in this state of affairs that underlies his hesitancy to bet the future on still unproved technologies. He isn't ready to turn his back on the ways and means of nature when efforts to improve the seedless watermelon are involved.

This is where O. J. Eigsti is as we approach the close of the twentieth century. I do not have license to say where science really is or where it is going. But it is easy to say that this one scientist has earned the right to profess his notion of the guideposts that are most likely to make science, nature, and humanity mutually beneficial. The guideposts he would construct to point the way into the new millennium ask to be examined against those that seem to be taking shape today, particularly where science, agriculture, and business came together.

What I allude to here is documented very well by Patent Number 5,007,198 that Dennis J. Gray and Gary W. Elmstrom applied for in 1998 and were granted in 1991. With the title "Process for the Accelerated Production of Triploid Seed for Seedless Watermelon Cultivars," this patent is a wonderfully informative window through which to view many aspects of where science is as the twentieth century fades into the twenty-first. The abstract tells us of "a novel process which facilitates the rapid and economical production of seedless watermelon seed." Featured in the process is the use of self-sterile tetraploid parental lines in the production of triploid seed.

Earlier conversations with O. J. assure us that he is quite familiar with the claims covered by this patent and the evidence that was submitted to support its granting. But not until I read the full text of the patent did I understand why his reactions were so emphatic. Of special note was the scientific principles and technological practices that underlie the production of seedless melon seed as they have been developed by O. J. and his colleagues. Yet there is no reference to work similar to that covered by the patent that was completed in China and elsewhere as early as 1980 and fully reported in the literature.

Central to the process for which this patent was granted is tissue culture, a laboratory procedure first developed by the American zoologist Ross Harrison in 1907. It was Alexis Carrel, however, who gained much publicity by keeping tissue from a chicken's heart alive indefinitely. Early in the 1930s, the American botanist Phillip R. White and the French botanist Roger Jean Gautheret adapted the process to enable them to study the onset of tumors in plant tissues. In 1960, Georges Morel, also French, improved the process by adding the essential hormones to the culture medium. This enabled him to culture thousands of orchid plants.

The tissue culture procedure is based on the fantastic ability of life processes to continue in isolated bits of tissue as long as they are maintained in an environment that meets all of the needs of the tissue while it was a part of the organism from which it was removed. To add to the potentialities of this procedure in the world of plants, each cell in the growth tissues of a plant, if properly cared for and nurtured, is "programmed" to develop all of the tissues and organs of a fully developed specimen of its parent species.

The details of the provision, "if properly cared for and nurtured," are spelled out in the patent.

> Tissue is taken directly from shoot or axillary bud apical meristems of growing plants . . . sterilized for 3 minutes in a 25% solution of commercial bleach, rinsed in sterile water, and the apical meristems micro-dissected. Apical meristems are placed on a solidified culture medium with growth regulators that encourage adventitious buds to form. . . After four weeks, cultures are transferred to fresh shoot elongation medium. . . .Numerous shoots grow from each original meristem. . .For culture increase, tissue masses on bud initiation medium are subdivided into several pieces every four weeks and transferred to fresh medium. Each original meristem is thereby induced to form a nearly unlimited number of buds. Cultures can be maintained and increased in this manner for over two years. . .
>
> For plant recovery, shoots approximately 2 cm long are cut and placed in rooting medium. . . When roots appear, the platelets are acclimated in soil in a humid covered chamber and kept for one week under artificial lights. . .After one week in the humid covered chamber under artificial lights, the chamber is then placed in a greenhouse. When new shoot growth is observed, the lid is gradually removed during the course of one week. The resulting plants are then transplanted to the field. Subsequent maintenance procedures are identical to those used for seedling plants.

Both the underlying rationale and the argument proclaiming the superiority of this procedure are being cited as evidence of evolutionary changes in what it has meant to be an agricultural scientist during the twentieth century. For changes in spirit, I focus on such statements as: "Although seedless watermelon cultivars [varieties] have been available for several years, technical difficulties in producing seed have kept the availability of planting stock low and the price of seed prohibitively high. Additionally, much variation in quality occurs between seedless watermelons. . . . The extremely slow accumulation of parental seed has been a major roadblock to the commercial production of triploid seed for seedless watermelons. In fact, this roadblock has been of such magnitude as to squelch development of a substantial seedless watermelon industry."

It isn't only that these statements are questionable. It is their lack of empathy, the absence of cooperation such as that marked O. J.'s relationship with Kihara that is missing in these statements. So is the assurance on the part of such early growers as the Larsons that seedless melons would be ultimately successful. Nor is it that there are not groups working together for their mutual advantage today. But these statements do bring to mind a sense of a dog-eat-dog attitude that O. J. did not encounter earlier in the century, as do statements to the effect that all a competitor needs to get started in the seedless watermelon business is to obtain one specimen from another watermelon breeder's tetraploid seed line and save its seeds.

Another putdown shows through such statements as: "Tetraploid seed . . . has proven to be very difficult to produce in large, commercially useful quantities. This is largely due to the fact that tetraploids exhibit a high degree of self-sterility. As a result of this sterility, very few melons develop in a field of tetraploid plants. Also, none or only a small number of seeds are usually produced in each self-pollinated melon. A productive cross will yield only 50–100 seeds per fruit. Therefore, ten or more years are typically required to increase seed of a new tetraploid line to commercially acceptable numbers." As before, these statements are not totally untrue, but seem to reflect judgment calls or the use of exaggeration to put a special spin on the messages they convey.

Much is made of the proclaimed sterility of tetraploid melons. O. J. questions these claims. He sees the production of normal, healthy pollen grains and ovules as being essential to the ultimate production of the triploid seeds that produce the seedless melons. His reasoning focuses on knowing that nature does not give much maneuvering space where pollination and fertilization occur in the normal process of getting a new watermelon on its way. As explained earlier, the trick is to encourage the development of the vehicle that is to carry the next generation embryo, but to leave the "baby carriage" unoccupied. To produce a high-quality seedless melon calls for enabling the former requirement to be met without restriction or interference. O. J. argues that this is best achieved by letting bees do their thing in the rows of tetraploid seed-line melons growing in their natural habitat.

It is the patenting of new developments that marks the most deeply disturbing changes in what it has meant to be an agricultural or horticultural scientist during the twentieth century. As far as I can determine, O. J. and his contemporaries gave little thought to patenting in all of the years they were struggling to make seedless watermelons commercially successful. Protecting "trade secrets" was about the only reason for pulling down the curtain on the open exchange of information. O. J. was chided by his friends more often for sharing his solutions to sticky problems than for being closed mouth about them.

This attitude is well illustrated in a paper O. J. presented to his compatriots in the seedless watermelon industry. In the section titled, "Seed Failures and Germination Difficulties," we read,

> Triploid seed quickly gained a notorious reputation for poor germination. This feature caused concern to the seed trade whose commodity, seed, must grow and produce plants if their business is to survive.
>
> Triploid embryos inside tetraploid seed coats show a great range in size and weight. Some seeds have embryos with only a microscopic mass of cells. Others have a large well developed embryo. Since the weight of seed coat is constant and the weight of the triploid embryo coat is not, it becomes possible to select for embryo size by differences in specific gravity. Appropriate machines are capable of separating the lighter weight, smaller embryos from the heavier weight, larger embryos. It is also noted that seed germination and seedling vigor are directly correlated to the larger embryo triploid seeds. The elimination of low specific gravity seed was a very important solution to a very difficult problem.
>
> A second cause for seedling failures is also related to the nature of embryo size in tetraploids. Between the tetraploid seed coat and the triploid embryo there usually occurs a large cavity not filled by endosperm or seed tissue. If the seed coats absorb water too quickly, the excess water accumulation does in some manner kill the embryo. Over-watering at the time of planting is the cause of many failures.
>
> The third problem associated with embryo-seed relations is the technique for seed separation from tetraploid flesh. Frequently when the watermelon pulp is fermented as a means of seed separation, there is no apparent harm done by pulp fermentation. However, triploid seed when allowed to remain in fermenting pulp, is damaged resulting in decreased rate of germination. This problem was solved by developing seed separation based on large volumes of water under a pressure of 1000 gpm.

If Patent Number 5,007,198 is to point the way into the new millennium, it puts a damper on the open exchange of successes and failures among research people during the early stages of discovery and invention. Consider the steps that are involved in this "claimed invention."

1. Produce and identify a suitable tetraploid parental cultivar
2. Clone the cultivar using unique tissue culture techniques
3. Transplant and grow the new plants in a suitable soil medium
4. Plant the tetraploid plants in a field or greenhouse in a mixed stand with diploid plants;
5. Collect the triploid seed from the fruit produced when the tetraploid plants are pollinated by the diploid plants.

People with only limited knowledge of damages levied against infringement of patents are concerned about the scope of this patent—not only in the seedless watermelon industry, but agricultural and horticultural research and development in general. All of which is to say that it is the "what's next?" portion of this book that stands to be most affected by the letter and, especially, the spirit of Patent Number 5,007,198.

Chapter 12

Awards and Recognition

THE AWARDS AND recognitions Dr. O. J. Eigsti has received tell us much about who he is, what his goals are, and what he stands for. Equally revealing are his attitudes toward the circumstances that determine the why and when of being chosen for recognition. A representative example is the Outstanding Service Award presented by the National Watermelon Growers Association at the annual meeting in Nashville, Tennessee, in 1996.

> *In 1971, they invited me to be on the program at New Orleans. I was to talk on seedless watermelons. And that's the last thing these watermelon growers wanted to hear. Boy, was there ever antagonism toward me. I had just received this picture from the cover of the Horticulture Society's publication. A beautiful full-color photo of a perfect seedless watermelon. I had enough copies to hand them to everyone. They looked at this picture and thought this guy's a fake. I gave my lecture. It was a rough go. The general feeling was that we didn't need seedless and we don't want seedless. I remembered the situation with hybrid corn. No one wanted those "nubbins" during the '20s and '30s. Then by 1935, everyone wanted hybrid corn and by 1940 the shift to hybrid was practically complete.*
>
> *What a difference at the Nashville meeting! The theme of all the exhibits was seedless watermelons. Some 25 percent of the melon production in the U.S. today is seedless. In Israel it is 50 percent. In 1989, when Karl and I spent two weeks in Israel, I was regarded as a celebrity, but not in my hometown. What do they say about the prophet*

in his hometown? What a change when I was called to the speaker's platform to receive that Outstanding Service Award.

O. J.'s awards and recognitions will continue. For now, thanks to one of his friends of many years, Dr. Max Britton, here is an example of the documentation that supports these honors—an example that nicely avoids the suggestion of obituary that so often spoils the spirit of such presentations.

> Since Dr. Orie J. Eigsti and I are both botanists, were members of the same academic department, and became very close friends, I have reason to know well his background. I enjoyed the opportunity to observe his successes in teaching and in research while we were faculty associates and to follow his subsequent professional life to date. His career has been noteworthy for both his dedication to pure science and to a parallel interest in the application of scientific research to the development of crop plants of economic significance.
>
> I believe that growing up on a successful Illinois farm in the heart of one of the world's richest agricultural areas, coupled with being the son of a banker, very probably brought him many insights into the importance of crop improvement and to the economics of living a good life. Such attributes, coupled with his inherent talent for science teaching and research, led him irrevocably to serve both arms of the research and development couplet. This pattern persists deep into his retirement. When I visited his experimental farm in Indiana during the past summer [1992] he was heavily engaged in basic research on the resistance of various tetraploid watermelons to several races of Fusarium wilt while, at the same time, entrepreneurs in other parts of the country were reaping and sharing rewards for his development of valuable seedless watermelon hybrids. These developments, of course, resulted from his earlier basic research on chromosomes.
>
> In 1935 Eigsti completed his doctoral research on the significance of chromosome morphology and chromosome set number (ploidy) to phyletic relationships of several species within one plant species. Armed with knowledge of the importance of polyploids to speculation in both native populations and agricultural cultivars he was very responsive in February 1937 to newly reported animal tissue and cellular reactions to a mititotic poison, colchicine. Realizing the implication to polyploidy, he resolved to determine if the chemical similarly affected plants. Although working on other research projects as an employee of the Carnegie Institution of

Washington at Cold Spring Harbor, he immediately began experiments, at night on his own time, which, within three days, demonstrated colchicine to be a powerful tool in creating polyploids. He was one of three or four workers to independently make this discovery in 1937. His first results were announced and an abstract published by others that year at a genetics meeting in France but he did not achieve publication under his own name until the following year.

From this time onward his name was inextricably linked to colchicine research in general and to polyploidy in plants specifically. His research continued to focus on both phyletic relationships and the development, breeding and maintenance of polyploid species, outstanding contributions being made to both of these research areas.

Another especially noteworthy achievement in 1937 was his development of a pollen tube methodology for chromosome study which was very important to cytotaxonomic investigations. A pollen tube contains only one cell, the generative cell, which undergoes any further division, and it has the haploid chromosome number. Briefly, culture media and environmental factors were elucidated and refined which permitted in vitro culturing and colchicine treatment directly on microscope slides. Under this treatment the chromosomes of only one haploid cell at a time are the object of examination and the chromosomes are mostly scattered in the pollen tube, thus making determination of chromosome numbers and morphology much simpler and more rapid. This technology marked a major technological development.

It was clear from his earliest work in 1937 that colchicine treatment provided the most efficient method of doubling chromosome numbers in diploid species of plants and, thus, the creation of new tetraploids. It was well known that tetraploids from whatever origin when crossed with diploids produced triploids which were always sterile. Eigsti became interested in watermelons since their triploids showed the inevitable sterility, the fruit was seedless, and the fruit also appeared to have many improved characteristics of potential interest to crop development. Through critical selection of diploid varieties of seed parents of known desirable traits, inducement of tetraploids by colchicine treatment, proper selection of diploid pollen parents, and by determination of suitable ratios of pollinators to tetraploids in the field, many promising triploids, the seedless hybrids, were successfully developed.

Growers of seedless watermelons, both in this country and Japan, soon discovered that among their presumed triploid populations, diploid seeded fruit often appeared and reached markets expecting seedless fruit. This disturbing condition increased in frequency in successive years in a given population, to the point that in a few generations tetraploids were no longer dominant. Some thought this problem arose as a result of environmental conditions. In 1957 Eigsti began research which was conclusively to prove the condition was genetic in origin. By 1961 he was able to report to a meeting of the American Institute of Horticultural Science at Purdue University (an eminent center of watermelon research), that the tetraploid populations were shifting to the diploid state through the process of parthenogenesis. Such diploids derived from tetraploids are referred to as polyhaploids. Polyhaploidy is a constant event among tetraploid populations propagated either by open or controlled pollination. Eigsti's discovery had both important theoretical and agricultural implications and was of enormous practical value to growers.

As Eigsti's research reputation and scientific stature thrived, he became a central figure on the forefront of research in his field. He was honored often by invitations to chair scientific meetings, organize and lead seminars and to give scholarly lectures. These responsibilities and the recognized need to keep in touch with all centers of related research, led him to extensive travel and often to extended stays in various parts of Europe, Asia and South America.

In the early forties, following a veritable explosion of colchicine-related research and publications, Eigsti recognized the need for a definitive work to trace the origins of colchicine research, summarize and interpret the accomplishments and to take stock of progress in the field. He began writing such a work in 1942 only to discover a few years later that a volume with similar objectives was in progress in Belgium. The two authors got together and in 1947 decided to join forces and develop a collaborative work. This collaboration between a plant scientist and a medical doctor brought unusual diversity and competency in the review and summarization of a very broad range of subject matter. The resulting volume, titled *Colchicine – in Agriculture, Medicine, Biology and Chemistry*, published in 1955, brought highly deserved credit to both authors.

Finally, another recognition of Eigsti's stature among botanical scientists was his role in the celebration of the Golden Jubilee of the Botanical Society of America. He was selected to prepare a paper on polyploidy. The volume of commemorative articles was published in 1957.

> From my point of view, my endorsement provides adequate assurance of meritorious original research on the part of Dr. Eigsti. With that matter at rest, I should like to add a matter which is of related interest in that it speaks to Eigsti as a whole man. More than forty years ago he foresaw a significant niche for seedless watermelons in the total spectrum of watermelon production. Knowing that every planting would require availability of a reliable, annual source of triploid seeds and that the production of such seeds was likely to remain a highly specialized process, he decided to attempt to provide that service. Toward this end, he organized a small family corporation which was named the American Seedless Watermelon Seed Corporation. I do not know the measure of success of this privately owned company but I do know the business was kept afloat and seeds of some very successful triploids were annually on the market.
>
> Some five or six years ago [1987–88] entrepreneurial interests large [sic] in crop production, including seed production and marketing, became interested in Eigsti's small business and entered into partnership arrangements which have quickly become highly productive and rewarding although, again, details are not known. To me, this little success story gives added luster and new dimensions to an already successful scientist.

It is the breadth of O. J. s recognitions and awards that tell us the most about what he has achieved and the kind of person he is. There was the Dustin Medal awarded by the University of Brussels in 1949 and the life membership in the Phytomorphology, Botanical Society of India that was presented in 1953. A luncheon to honor the "developer" of the seedless watermelon hosted by the Munday (Texas) Chamber of Commerce is remembered by O. J. as providing almost equal satisfaction. Obviously, seedless watermelon was served at this luncheon.

O. J. received the first Industry Division President's Corporate Council Award of the Industrial Division of the American Society of Horticultural Science at a joint session with the Canadian Industrial Division in 1995. This award honors "the outstanding contributions of an industry horticulturist to the science and profession of horticulture, as evidenced by an original research project method, invention, or other innovative application of horticultural science in the private sector. This award, together with similar recognition by the Indiana Watermelon Association, represent the satisfactions that served to balance the frustrations that have been a part of O. J.'s lifelong compulsion to make the seedless watermelon a commercial success.

Glossary

Acquired character— A feature or behavior that a parent can hand down to offspring that was acquired by something the parent experienced rather than by way of the parent's genetic inheritance. A very controversial term. At one extreme of belief, pierce the ears of enough generations of daughters and, eventually, girls will be born with pierced ears. At a much lower level of belief, after enough generations of bluebirds have built the same kind of nest, eventually new generations of incubator hatched and reared bluebirds will build the same kind of nest. Disbelievers allow no newly acquired trait, physical or behaviorial, to be handed down unless a gene has been reconstructed.

Albinism— A trait that does not permit pigmentation. Usually the effect of recessive genes that control the production of melanin.

Alteration of generations— Action that marks the evolution of plants from primarily vegetative to sexual forms of reproduction, for example, from the production of spores as in puffballs to seeds as in watermelons.

Anther— The part of the flower where the pollen appears.

Apical meristem— Tissue at the growing portions of a plant where the cells contain a complete set of that plant's genes.

Aquifer— A layer of porous rock or gravel thick enough to hold large quantities of water and near enough to the surface to supply water for homes or irrigation projects.

Aseptic— Free from disease causing organisms.

Bacteriophage— A virus that is capable of rearranging DNA molecules particularly in bacteria.

Breed— Animals that share similar characteristics, e. g., cocker spaniels or black angus cattle. Equivalent to variety or cultivar in plants. e.g., the Tri-X313 watermelon.

Buoyancy of hybridzation— An early term for the still mysterious enhanced vigor shown by the offspring from crossing parent lines in which there had been no cross ferdlization for several generations.

Chromosome— Coming together of genes in a specific order during cell division. Named as they are because they are uniquely sensitive to certain dyes or stains.

Chromosome mapping— Determining the location on a specific chromosome of a gene that is known to control a specific trait.

Clones— Individuals of the same species or variety with identical genomes, that is, identical genes.

Colchicine— One of nature's chemicals that, among other properties, can cause daughter cells to have two or more times the usual number of sets of chromosomes.

Colchicine mitosis— Cell division carried out under the influence of colchicine.

Cross fertilization— Arranging to have the egg cells produced by one plant to be fertilized by sperm cells produced by a plant of a different variety.

Cultivar— A plant group within the same species whose members share similar characteristics. Equivalent to variety.

Culture medium— A preparation in which the ingredients are known to permit certain organisms to grow under controlled conditions.

Cytological— Studying life's events and circumstances at the cell level.

Cytoplasm— Everything inside a cell except the nucleus and its contents.

Daughter cells— A convenient way to refer to the results from mitosis.

Diploid— The condition where cells contain a single set of chromosomes in which each chromosome is duplicated.

Disease resistance— Inherited immunity in which the effects from one or more genes protect against specific disease-causing organisms.

DNA— Deoxyribose nucleic acid. The "stuff" of genes from which are constructed the "informational molecules" that are what heredity is all about.

Dust storms— geological catastrophe that peaked during the 1930s and led to new soil conservation and protection practices.

Embryo— A new plant or animal during its very early development. Plants at the embryo stage of growth are usually protected and nourished within the seed. A common name is germ as in wheat germ.

Endosperm— Nutritive tissues within the seed that nourish the embryo. These tissues also indirectly nourish the animal kingdom including, of course, humanity.

Eugenics— The branch of science that is concerned with the role genetics might play in furthering the well-being of humanity.

Factor— Equivalent to and eventually replaced by the term, gene.

Gametophyte— The stage in the life cycle of a plant when egg and sperm cells are produced, that is, the gametes.

Gamma rays— The relatively low energy but highly penetrating form of nuclear radiations.

Gene— Perhaps best described as one of the "information molecules" that are invariably present whenever characteristics are handed down from parents to their offspring.

Generative cell— Any cell in an organism whose function relates primarily to the continuation of the species. More specifically, the nucleus in the pollen tube that produces sperm cells.

Genetic diversity— The concept that a species population stands a better chance of surviving epidemic or environmental threats if the full array of the species genes is maintained within the population. A population that consists of a single clone is considered to be at risk.

Genetic engineering— Transferring a gene from one species to another where it functions as it does in its original species. Also known as recombinant DNA.

Genome— All of the genes or all of the positions for genes on all of an individual's sum total of chromosomes.

Germination— When a "resting" or dormant seed or spore resumes active growth.

Germ plasm— A somewhat abandoned term that separates those cells that are primarily involved in heredity from all other "body" cells.

Glacier— One of the enormous ice sheets that crept from the Arctic and constructed and reconstructed North America including much of the "Corn Belt."

Grafting— Transferring a twig, (scion), or bud from an individual of one variety to the root or stem stock of another variety in such a way as to insure that the growth tissues of both are in touch.

Great Depression— The period beginning in 1929 when the almost total collapse of the financial world created conditions in which money problems became uppermost even to the point of putting sheer survival at risk for all but a very few.

Haploid— Half the usual number of chromosomes; as in egg or sperm.

Herbicide— Weed killer.

Hectare— Equivalent to approximately 2.47 acres.

Heredity— The handing down of characteristics, both form and function, from parents to their offspring.

Heterosis— Equivalent to hybrid vigor, the condition that comes with crossing two different lines of inbred parents.

Hexaploid— Six times the usual number of pairs of chromosomes.

Homozygosity— A minimum number of unlike genes, (alleles), for the same trait, e. g., pure dominance or pure recessive.

Hormone— A substance produced, perhaps, in one part of the body and transported elsewhere and uniquely effective in influencing a specific body chemistry.

Horticulture— The branch of agriculture that features fruits, vegetables, and ornamentals.

Hybrid— Offspring resulting from crossing parents who differ in at least one characteristic.

Hybrid vigor— The mysterious increase in the magnitude or intensity of certain traits that accompanies hybridization. Equivalent to heterosis.

Information molecules— The name for genes that focuses more on function than on their nature.

Inhibitor gene— A gene that overpowers or inhibits the action of another gene.

Laws of Learning— Based on the belief that there are certain actions and conditions that are essential to and, in fact, are sufficient to insure success in the classroom.

Lysenkoism— Promoting the validity of inheritance of acquired characteristics as infrastructure to support the Communistic form of government.

Male sterility— A condition that results from the action of certain genes located either on chromosomes or in the cytoplasm.

Meiosis— A form of cell division giving rise to egg and sperm cells in which each daughter cell receives only half the number of chromosomes that exist in the body cells of that species.

Metaphase— The interval during mitosis when the chromosomes are lined up across the middle of the nucleus.

Mitosis— The cell division process that accompanies the replication of each "information molecule."

Morphology— The branch of science that focuses primarily on the kinds of organs that distinguish one organism from another and how these organs develop.

Mutation— The observable effects resulting from an apparent change in the atomic make-up of an "informational molecule," that is, a gene.

Nucleus— A relatively denser body within cells that serves as the home base for mitosis and meiosis.

Oncogenes— A late 20th Century term describing certain genes that influence or control the actions of other genes.

Ovary— The organ in which egg cells develop. In plants, an enlarged ovary becomes the fruit.

Ovule— A developing seed containing a not yet fertilized egg cell.

Ovum— An egg cell.

Parthenogenesis— The development of a gamete, usually an ovum, into a new individual without being fertilized, that is, interacting with a gamete of the opposite sex.

Penny Auction— A relatively rare practice durng the Great Depression whereby neighbors protected each other against having their property repossessed by mortgage holders.

Phage— Equivalent to bacteriophage, a virus infecting bacteria and capable of rearranging the atomic make-up of the bacteria's DNA.

Phyletic— Dealing with determining an organism's proper place in the classification scheme.

Pollen tube— After pollen finds it way to the stigma of a flower, a pollen tube gradually works its way down the style enabling the chromosomes of the sperm cell to reach and, eventually, to fertilize the egg cell in the ovary.

Pollenation—Pollen grains being transferred from anthers to stigmas.

Polyploidy— Having more than the usual number, (diploid), of sets of chromosomes.

Polyhaploids— Egg or sperm cells that contain more than the usual halfnumber of sets of chromosomes.

Post-Sputnik— A significant reform era in the evolution of science education in the U.S. that developed in response to Russia being the first nation to successfully launch a satellite.

Prophase— The time interval during mitosis or meiosis when the genes and other chromatin materials are coming together to form clearly distinguishable cbromosomes.

Recombinant DNA— The technological process in which the genome of a species is reconstructed, especially by introducing one or more genes from another species. Equivalent to genetic engineering.

Reduction phase— The interval during meiosis when each daughter cell receives only half the number of sets of chromosomes that would have been transferred dunng mitosis.

Scion— During grafting, the twig that is transferred to the stock.

Selection— The process or state of affairs in which only certain offspring are permitted to give rise to the next generation. Selection processes may be natural as "survival of the fittest" that results from the high reproductive rate of most species, or technological as in saving for seed the offspring that best conforms to a plant or animal breeder's ideal variety or breed.

Species— A group of plants or animals whose members can interbreed and their offspring are equally capable of having offspring.

Sperm— The male reproductive cell.

Spindle fibers— Almost invisible threads that seem to control chromosomes as they move toward their final position during mitosis or meiosis.

Sport— An offspring that differs markedly from its parents and can transmit the difference(s) to its offspring. The result of a change in the atomic structure of a gene, hence, a mutation.

Stigma— The part of a flower that receives pollen.

Stock— During grafting, the stem or root that receives the scion or bud.

Survival of the fittest— A Darwinian term that derives from the high reproductive rates of species with the result that all offspring simply cannot be supported by the environment.

Tetraploid— Possessing four times the usual number of sets of chromosomes.

Tissue culture— Maintaining small groups of cells from plant or animal tissues under conditions that permit continued growth.

Trace element— Any one of several chemical elements that are essential to plant growth but in very low amounts.

Transgenic— A plant or animal individual or species whose ancestry traces to being a product of recombinant DNA, that is, to acquiring one or more genes from another variety or species.

Translocation— The transfer of a gene during recombinant DNA.

Triploid— Three times the usual number of chromosomes.

Unit character— A term that is equivalent to but in use before the term, gene, was established.

Variety— A subdivision of a species whose members share certain distinguishing characteristics. Equivalent to cultivar.

Sources

Andrus, C. F., and others. "Production of Seedless Watermelons." U.S. Department of Agriculture Vegetable Breeding Laboratory.

Bailey, L. H., and William M. Coleman. *First Course in Biology*. New York: Macmillan, 1908.

Britton, Maxwell Edwin. Personal correspondence.

Caldwell, Otis, and William Lewis Eikenberry. *Elements of General Science With Experiments*. Ginn and Company, 1926.

Clark, Bertha M. *New Introduction To Science*. New York: American Book Company, 1928.

Cross, George Lynn. *The University of Oklahoma Research Institute*. Norman: University of Oklahoma, 1986.

Duckworth, Donna. "History and Basic Properties of Bacterial Viruses." *Phage Technology*, 1987.

Eigsti, O. J. "A Comparative Study of Male Gametophytes in Tradescantia reflexia Ref. and Lilium Canadense L." *American Journal of Botany*, 1937.

Eigsti, O. J. "A Contribution To the Life History of Reseda odorata" *MS Thesis*. University of Illinois, 1933.

Eigsti, O. J. "A Morphological and Cytological Investigation of Certain Members of Resedaceae." *Abstract of Ph.D. thesis*. University of Illinois, 1935.

Eigsti. O. J. "The Effects of Colchicine Upon the Division of the Generative Cell in Polygonatum, Tradescantia, and Lilium." *American Journal of Botany*, 27: 512-524. 1924.

Eigsti, O. J. "Induced Polyploidy." *Ibid*, 44: 272-279. (This paper also appeared in Steere, W. C., *50 Years of Botany, Golden Jubilee Volume of the Botanical Society of America*. New York: McGraw-Hill, 1958.)

Eigsti. O. J. and Pierre Dustin. *Colcicine – In Agriculture, Medicine, Biology, and Chemistry*. Ames: The Iowa State College Press, 1955.

Eigsti, O. J. "A Cytological Study of Colchicine Effects in the Induction of Polyploidy in Plants." *Proceedings of the National Academy of Science*, 24: 56-63. 1938.

Encyclopedia of Agriculture.

Glordan, Judith C. "Science in the 21st Century: Harmony for the Millennium." *Chemical and Engineenng News*, December 2, 1996. p. 32.

Gray, Dennis J., and Gary W. Elmstrom. "Process for the Accelerated Production of Triploid Seeds for Seedless Watermelon Cultivars." U. S. Patent Number 5,007,198. 1991.

Hamilton, Carl. In No Time at All. Ames: The Iowa State University Press, 1974.

Harold, Franklin M. "From Morphogenes to Morphogenesis." *Microbiology*, 141, 2765-2778, 1995.

Harwood, W. S. *New Creations In Plant Life*. New York: Macmillan, 1905.

Hildebrand, John. *Mapping the Farm*. New York: Alfred Knopf, 1995.

Hunter, George William. *A Civic Biology Presented in Problems*. New York: Macmillan, 1914.

Jordan, David Starr, and Vernon Lyman Kelley. *The Scientific Aspects of Luther Burbank's Work*. San Francisco: Philopolis Press, 1908.

Keller, Evelyn Fox. *A Feeling for the Organism: The Life and Work of Barbara McClintock*. San Francisco: W. H. Freeman and Company, 1983,

Kihara, Hitoshi. "Triploid Watermelons." *Proceedings of the American Society of Horticultural Science*, 58: 217-30,1952.

Klinker, J. Edward. "Experimental Production of Seedless Watermelons." *Progress Report 15,* October, 1953. Kentucky Experiment Station, Lexington, KY.

Lysenko, L. Y. *The Situation in Biological Science: Proceedings of the Lenin Academy of Agricultural Sciences of the U.S.S.R: Session July 31 - August 7, 1948*. Moscow: Foreign Languages Publishing House, 1949.

Mendel, Gregor. "Experiments in Plant Hybridization." Originally published in *Verhandlungen Naturforschender Verein* in 1865. Translation appears in *The Principles of Genetics.* Edmund W. Sinnott, L. G. Dunn, and Theodosius Dobzhansky. New York: McGraw-Hill, 1953.

Miller,Edwin C. *Plant Physiology*. New York: McGraw-Hill, 1931.

Ming, Wang, Zing Xingping, and Zhang Xian, "Breeding Few-seed Seedless Watermelon Via Chromosome Reciprocal Translation Induces by Gamma Ray" *CGC*: 60(1988).

Rhodes, William. Personal correspondence.

Sinnott, Edmund W., L. C. Dunn, and Theodosius Dobzhansky. *Principles of Genetics*. New York: McGraw-Hill, 1955.

Slye, Maud. "Cancer Heredity," *Journal of the American Medical Association*, 85: October 25, 1925.

Soth, Lauren. *Farm Troubles*. Princeton, New Jersey: Princeton University Press, 1957.

Strauss, William, and Neil Howe. *Generations: The History of America's Future*. New York: William Morrow, 1991.

Williamson, L. A., and G. A. Rice. *Principles of Secondary Education*. Boston: Ginn and Company, 1927.

Wong, Cheong-Yin. "Induced Parthenocarpy of Watermelon, Cucumber and Pepper." *Science*. 89: No. 2314, p. 407-418, 1930.

Index